CULTURAL GEOGRAPHY

Cultural Geography

R.K. Tripathi

Commonwealth

Cultural Geography

First Published, 2013

ISBN 81-7169-640-6

Published by :

Ajay Verma
For Commonwealth Publishers
4831/24, Prahlad Street,
Ansari Road, Darya Ganj,
New Delhi - 110002
Ph. : 3272541, 3257835
Fax : 91- 011- 3257835
e-mail : campusbooks@hotmail.com

Typesetting by :
Arjun Computer's
Delhi - 110051

Printed at :
Roshan Offset Printers
Delhi - 53 Ph. : 2910931

Preface

Cultural geography is, and always has been the study of peoples and their works, the site, situation and specific time. Careful attention is given to the areal distribution of the various factors. Since the time factor is of great importance, human geography is just as concerned with the past as with the present.

Today human geography is still undergoing some changes, but the hard core of the subject is well established. The most noteworthy changes are in techniques and emphasis rather than in basic content. A high degree of specialisation is also apparent, in both method and content. Some workers depend mostly on statistics; others concentrate on aerial photographs.

There are fewer studies which attempt a complete, overall coverage of human geography. Each special study, however contributes something to the main goal of cultural geography: the description and explanation of the great variety of peoples and cultures, as they exist in highly diversified environments.

Editor

Contents

1
Land Use and Conservation

The problems of conservation are most pressing in regions of dense population and highly developed machine technology, but this is obviously an over-simplification. In some areas of dense population most of the problems of conservation have been solved, for the time being at least, while in some sparsely peopled areas the resources are being rapidly exhausted.

In any broad view of conservation and land use, one is impressed with three aspects: the variety of resources in different parts of the earth; the diversification of peoples, customs, and land use; and the present rapid intensification of land use as population increases and machine technology spreads. If one considers the variety of resources, on the one hand, and the various types of land use, on the other, there is the temptation to infer a simple correlation between them.

It would simplify the problems of defining land-use types, for example, if they could be correlated with climate, soil, natural vegetation, and mineral deposits. To be sure, a rough correlation can be made between agricultural types and climate. It is possible to speak of the "wheat climate," the "corn climate," and the "coffee climate" and to devise climatic definitions to fit the major distribution of these crops. But it must be noted that wheat grows better in the corn climate than it does in the wheat climate and that all three crops grow fairly well in a variety of climates. Land use certainly depends on physical resources, but it also depends on the character of the people and the stage of their development.

As life becomes more diverse and intense, the attitude of man toward the resources of the earth-toward conservation-changes, sometimes very abruptly. In the early days of the hunting-gathering-fishing peoples the idea of conservation, of saving resources, had not emerged. They collected what they could, usually fruits and nuts.

If nature was kind, there was an abundance of everything; if disaster struck—drought, flood, or fire—the people probably blamed it on the "spirits." The attitude of the modem American pioneers was scarcely better, and the effect was much worse. They put the axe and fire to forests to clear land. In eastern North America oak trees, which would be worth $1,000 each today, were cut and burned to clear fields for corn and tobacco. These pioneers could not foresee, with a whole continent before them to exploit, that in a few generations all the good land would be occupied and most of the wood forests destroyed. In some parts of the world the attitude of pioneers still prevails, but many peoples have developed a different point of view. Thomas Jefferson and his contemporaries in the Virginia piedmont learned to plant trees, to terrace the hill slopes, to cultivate carefully, and to improve the soil by the use of fertilizers and manures.

The intensification of life involves more people, with increasing desires, more tools, more trade, more communication, and a greater transfer of goods and ideas. The basic needs of man—air, food, and water—are almost forgotten in the demand for the perquisites of modern civilization. The demand for such things as elaborate houses, automobiles, airplanes, and space vehicles has greatly intensified production and exploitation of the earth. Strangely enough, the intensification plus improved circulation is tending to decrease the diversification in man's activities. Marine farming with its large tractors, plows, and harvesters, once limited to very large farms, has been adapted to smaller ones. War has accelerated this intensification and has led to more uniformity in production methods. During World War II the United States shipped tractors to England and the Soviet Union in order to make up for their loss of manpower in military service.

Hunting, gathering, and fishing

The hunting, gathering, and fishing economy, once widespread over many parts of the earth, is now confined, for the most part, to the fringes of the Arctic Ocean in northern North America and northern Eurasia. To be sure, many other people practice some hunting, gathering, and fishing along with agriculture as a means of subsistence, but in the arctic regions agriculture is impossible and the people must depend entirely upon hunting, gathering, and fishing. This is mainly a subsistence form of land use and the natives, whether they are Eskimos, Lapps, Tungus, or Yakuts, must use a great deal of skill an invention in order to live in the very harsh, difficult climate. The reindeer, a browsing animal, enables some of these peoples to live a partially pastoral life, in contrast to the pure hunters, gatherers, and fishers.

All the arctic peoples depend upon wild animals to a considerable extent for their food, clothing, and shelter since vegetation is very scarce and very limited in variety. Some over-browsing has undoubtedly occurred, and the numbers of native animals have been reduced. However, the population is so sparse and the area so vast that man-made changes are scarcely noticed.

The population has been kept small by the hazards of the area—cold, starvation, and disease. Diseases introduced by the white man have been especially devastating. These regions are definitely on the fringe of human occupation and up to the present have had very little population pressure on them.

Forest production

Forest production in the world today may be divided into tropical and temperate types. The tropical forests, located mostly in central South America, equatorial Africa, and Indonesia, have a great many kinds of trees, mostly hardwood. Their very variety makes it difficult to harvest them, since any one wood, such as mahogany, is represented only by scattered trees throughout a large area. Little attention has been given to conservation of tropical forests, perhaps because the exploitation is usually on a small scale and regrowth is rapid. A few species, however, such as teak and some varieties of mahogany are becoming scarce.

The northern zones, in which forest production is dominant, W also utilized in other ways. Many places have mineral deposits of considerable value; in some favourable locations agriculture is possible, especially where there is a good local market for agricultural products, perhaps adjacent to a large lumber or paper mill. In the last half century, great attention has been given to conservation of the forests in various countries, after the complete destruction of large areas of forest land, particularly by clear-cutting operations. Norway, Sweden, and parts of Finland the coniferous forest has been largely cut-over, and the logging industry is now carefully regulated, whether on private or state-owned lands. The excellent pine forests of Michigan and Minnesota have been almost completely destroyed and in such a manner that they cannot grow back without expensive replanting.

A new approach that is taken in many places today is the tree farm, in which a large tract of timber is carefully managed, usually by a private company. Precautions are taken to prevent fires and to protect, the seeds from rodents, the young seedlings from animals, and the trees from diseases.

Logging of small areas is carefully supervised so as to permit rapid reseeding, prefer ably by natural means, from the nearby remaining ands. Usually, a tree farm includes many thousands of acres of forest land, and the company which owns it often controls many tree farms in the vicinity.

Mining

Mining is very widely distributed in various parts of the World, and there is a great variety in minerals. Mining is often referred to as the "robber economy'. Since every bit of mineral mined means that much less remaining for future generations. Some minerals are quite abundant; others are moderately abundant in some specific localities; still other minerals are scarce or expensive to move, and if they are badly needed in industry in a particular country, they are considered to be strategic minerals.

The strategic minerals present some special problems in conservation. Is it better for a country to subsidize its mining industry and produce some of the strategic minerals at high cost, or is it better to import these minerals and stockpile them for future use?

Another type of problem is the effect of mining on other forms of production. Placer mining and strip mining often destroy the soils and leave once fertile areas totally unfit for agriculture or grazing. Also, the fumes from copper smelters usually located near the mines often destroy all vegetation within a radius of several miles.

Nomadic herding

Over much of the earth's surface there is too little rain for cultivation of any kind except where special local water supplies are available for oases. These and lands, if occupied at all, are mostly used for pastoral purposes as areas for nomadic herding, which is a wandering, subsistence type of land use, or as commercial grazing areas, representing the commercial ranch type. Strictly speaking, neither of these types is agricultural but they are commonly associated with agriculture and, in some areas, there is some cultivation and cropping.

The occupation of the arid and semiarid land by nomadic peoples has brought about some is very uneven; many parts are completely unpopulated, mostly because there is no pasture or water for the herds and flocks.

Although the character of the peoples and their nomadic life varies somewhat from place, to place, their mode of living is generally similar throughout this vast region. They are forced to migrate from time to time in order to find forage and water for their livestock.

Most of the nomadic groups have a variety of livestock horses, cattle, sheep, and goats, for example. In recent years the impact of neighbouring industrialized peoples has had a great effect upon the life of the nomads. In Algeria and southern U.S.S.R., for example, the nomads have more or less been forced to settle down and become agriculturalists because their best grazing lands have been taken over by other peoples for agriculture. The occupation of the arid and semiarid lands by nomadic peoples has brought about some significant changes in the natural vegetation.

The population tends to expand to the limit of the food supply. In times of excessive drought the native grasses and other forage plants are so reduced as to change the whole aspect of the land. When rains again occur, the original native grasses may be replaced by less desirable species. Once the pastures are badly damaged, the nomads tend to migrate to other regions and, in time, the damaged pastures are at least partially restored.

Commercial grazing

Commercial grazing is a fairly recent development. The chief regions are western United States, northern Mexico, and parts of Venezuela, Brazil, Argentina, South Africa, New Zealand, and Australia. Large areas of semiarid land, though too dry even for dry farming, provide a large quantity of forage for livestock, which can be marketed, in one form or another, in the great industrial regions far from the grazing lands.

The characteristic feature of commercial grazing is an established ranch or headquarters, where the livestock can be winterfed, branded, and prepared for the market. Some of the ranch lands are fenced, some are not; in any case, many hundreds or even thousands of acres of land are necessary for one ranch because, in the lands too dry for agriculture, the carrying capacity of the grasslands is so small that from 15 to 75 acres of pasture are needed for one cow.

In a few places, as in southwestern Texas, the goat is of considerable importance, but cattle, for beef and hides, and sheep, for mutton and wool, are the dominant animals in the areas of commercial grazing. In western United States the principal sheep areas tend to be concentrated near the mountains, which provide summer grazing for the animals, or near irrigated lands which supply winter feed.

Range sheep production, as contrasted to the sheep production on humid farms, has declined in recent years. In the more remote areas of commercial grazing, such as southern Argentina, the principal product is wool, since it can be exported at a profit even under difficult transportation conditions. More and more, the important areas of commercial grazing are establishing freezing plants so that the meat, mostly beef and mutton, may be shipped in a frozen state and thus bring a higher price in the world market.

Hoe culture

In hoe culture new land is cleared from time to time so as to maintain yield and fertility. It is common practice to girdle the trees by cutting through the bark or to burn the forest, following which crops are sown between the standing tree trunks. It is characteristic of hoe culture that fertilizer is not used. Hoe culture is practiced today on the fringes of plow culture, especially where steepness of slope or character of the soil makes plow culture unfeasible, as in tropical America, Africa, and southeastern Asia. Various crops are grown for food and fiber, including maize, bananas, manioc, yams, squash, and beans.

The native term applied to this form of cultivation is usually "shifting patch" cultivation. It is typical of hoe culture that, in addition to hand-implement cultivation, there is usually a periodic shifting of the patches, allowing a natural rotation to renew the nutrients in the soil. In some places the occupation is made more permanent by the construction of terraces on steep hillsides. The hoe-culture settlements are scattered, usually clearings in the forest. Occasionally some of the towns or villages are large, but the usual settlement probably contains not more than 200 or 300 people. Over the centuries hoe culture has resulted in some soil erosion in the most hazardous places and, undoubtedly, great changes have been made in the natural vegetation, but the changes are very slight compared to those brought about by plow culture.

Plow culture

The beginning of plow culture, a much more advanced form of cultivation than hoe culture, depended upon a combination of circumstances, particularly the invention of the plow and the domestication of herd animals large enough to pull a plow. Also associated with this beginning was the development of metallurgy to the point where a metallic plowshare could be invented. To be sure, wooden plows without metal are still used in isolated cases.

The earliest plow culture probably originated in China, India, and southwestern Asia and was associated with the cultivation of the small grains, particularly wheat, barley, and the millets. The peoples who cultivated these plants probably had sheep, goats, asses, horses, and cattle, but only the larger animals were suitable for pulling the plow. There is a great deal of physical diversity in the plow-culture areas; plow culture exists in a great variety of climate and terrain, from the irrigated oases around the fringes of the Sahara to the cool lands of northwestern Europe and the tropical lands of southeastern Asia.

The plow-culture peoples have a much greater variety of crops available for cultivation than the hoe-culture peoples. And, with the possibility of much more thorough and larger-scale cultivation, some of the crops have become very important: for example, grains in the western part of the plow-culture area, notably in the Mediterranean region and northwestern Europe.

The part that livestock plays in plow culture, aside from the mere drawing of the plow, varies with the nature of the people and also with the climate under which the crops and the livestock are grown. The following types of plow culture are based mainly on differences in crops and livestock, their uses, and the methods of production.

Rice farming

In southeastern Asia, including south China and India, rice farming is one of the most dominant forms of plow culture. This region is particularly suited to rice in that the summers are warm and rainy and, in many places, the winters are dry and mild.

Rice requires a long growing season with warm weather and abundant rain; it profits by a drier season near harvest time so that it may ripen and also be dried and preserved after harvesting. Rice is cultivated in two ways: it may be grown by broadcast seeding, particularly on hillsides after the land has been plowed and otherwise prepared; but the most important form of rice cultivation is by planting in small enclosures or paddies. In the regions of heavy rainfall it is only necessary to build up a small ridge to enclose the paddy—the rains flood the paddy from time to time, and by a little manipulation, the paddy can be drained. This method of alternate drainage and flooding yields the greatest returns; also, the rice fields can be used for other crops, such as wheat, barley, and millet, during the mild winter season.

Rice farming is a very intensive form of agriculture in which a large amount of labor is expended upon a small plot of land. Associated with rice farming is the cultivation of other crops—small grains, vegetables, spices, and many kinds of tree and bush crops—on the adjacent hillsides. In the areas of lighter, rainfall, irrigation is necessary to get the maximum yield of rice. This can often be done by diverting a stream and leading it onto the paddies, or by digging shallow wells in the alluvial soils of the great river flood plains, such as the Ganges or the lower Yangtze.

A number of animals, including horses and oxen, are used for draft purposes in the areas of rice farming, but the most successful draft animal for the rice paddies is the carabao (water buffalo). This animal likes the muddy, wet rice paddy and is able to pull a plow through it to better advantage than other animals. However, as noted above, the rototiller is displacing the draft animals in some regions.

Mediterranean agriculture

Certainly this climate is not especially suited to rice: rice can be grown, of course, in the dry summer with the addition of enormous amounts of irrigation water, but it will not grow well in winter when the rains are most abundant. On the other hand, wheat and barley,thrive in the mild winters, growing slowly through these long rainy periods and taking,advantage of most of the rain that falls.

Consequently, the combination of crops grown in these areas, to be sure, is very different from the crops grown in the rice farming regions. Great emphasis is placed upon the growing of winter grains with the addition of crops which can be grown in the summer, either with or without irrigation. In the latter category is the olive tree, which is characteristic of Mediterranean agriculture; with few exceptions, it grows only in the Mediterranean lands. It requires long, hot, dry summers, deriving moisture by means of long roots which reach down into the ground water. The olive makes up in many ways for the scarcity of animal products, taking the place of meat, milk, and cheese in the diet of the people, particularly in certain areas of the Old World. Olive oil is readily preserved and, therefore, can be carried over from one season to another. Conditions for livestock are not favourable. Forage is scarce and limited in quality; although the brushy areas of the Mediterranean region furnish forage for goats and ~asses, cattle and horses cannot thrive on it.

Other crops are the citrus fruits—orange, lemon, and citron—and the grape. Some of the citrus fruits mature principally in the winter and form a welcome supplement to the winter grains. The grape occupies a very special place because of its wine-making qualities.

Mediterranean agriculture has had contributions from a variety of places. Some of the contributions (citrus fruits, for example) came from the rice-growing regions of southeastern Asia. As time went on, there were contributions in the form of livestock from the nomadic peoples who lived on the fringe of the Mediterranean lands. In more recent times (after the Age of Discovery), a number of crops came from the New World, including maize, potatoes, cotton, and tomatoes; few of the New World crops have been successful on a large scale, however.

Like rice culture, Mediterranean agriculture involves a large amount of labor applied to small plots of land—a kind of garden culture. Cultivation is intensified by several devices, among which are multiple cropping (two or more crops on the same field), intercultivation. (alternate rows of different crops), and two-story cropping (in which grapes, for example, are grown on a supporting framework several feet above the ground while vegetables grow underneath).

Terracing with rock walls makes it possible to use steep hill slopes and, as long as the farms are small and labor is abundant, there is little or no erosion problem since the rains are usually light and gentle. But serious erosion has occurred on the hill slopes as a result of deforestation and fires. This is especially true on the Adriatic coast of Yugoslavia, where all the soil has been removed over large areas, exposing bare rock.

Mediterranean agriculture, which was largely for local subsistence in the initial stages of development, has become highly commercial in many regions, notably in California. Machine methods are now applied, and the crops are highly specialized. In most cases the principle of careful cultivation has continued, thanks perhaps to the migration of many thousands of people from the original Mediterranean lands to the newer districts in California, Australia, and South Africa. If the light rainfall minimizes soil erosion, it accentuates the water problem. For many cities in the Mediterranean lands, it is necessary to bring water from long distances.

Small-grain and livestock farming

North of the Mediterranean lands in Europe, people are faced with a different environment and have developed a more general type of farming, in which the field cultivation of grains and forage crops and the raising of livestock play the dominant role. This type of agriculture reaches from northern Spain through France, the Low Countries, Germany, and Poland to western U.S.S.R. Only small areas of this type are to be found outside of Europe, notably in New Zealand and the Great Lakes region of North America.

The winters are colder and, perhaps, of greater consequence to land use than they are in the Mediterranean lands; since the summers are much cooler, some of the crops of the Mediterranean lands cannot be grown. The olive cannot be grown since it does not thrive the rainy summers of central and northern Europe and since it is susceptible to heavy frost in any period of the year. The citrus fruits are even more restricted in their distribution than the olive. The grape is successfully cultivated in the southern part of central Europe, as far north as Paris and southern Germany.

But the small grains—wheat, barley, rye, and oats—thrive in the moist climates of central and northern Europe to even better advantage than in the

Mediterranean lands. Pastures are better, and hay is readily produced as feed for the livestock that are used as draft animals, for meat, and for dairy products. Forage crops are diversified as a result of the introduction of potatoes, sugar beets, mangels, and maize from the New World.

The most important of these is the white potato, which is used for food, for livestock feed, and for the manufacture of starch and alcohol. Four countries—France, Germany, Poland, and U.S.S.R.—are the greatest producers and consumers of potatoes. The introduction of the sugar beet was also important since Europe has no other satisfactory means of producing sugar. The only previous production of sugar was from honey and this only in small quantities. The growing of sugar beets makes it possible to produce quantities of sugar at low cost, rivalling the production of cane sugar in the tropical lands.

Another important root crop is the mangel, which grows well and is used mostly for livestock feed. Since a good part of this area is too far to the north to produce good wine grapes, there is emphasis upon barley and hops for beer and other malt beverages. One crop that cannot be successfully introduced a grain into northwestern Europe is maize since the growing season is too short and the high temperatures necessary to produce a good crop are lacking; maize can be own for fodder only. The small-grain and livestock type of farming, evolving slowly over the centuries, presents the best traditions in conservation.

As population increases, the pressure on the land leads to improved methods of cultivation and to the reclamation of salt marshes and inland freshwater swamps. By the application of mineral fertilizers and green-manure crops, and by long cultivation, the quality of soils has been improved. Most of the region lies in the great lowland plain of western and central Europe, but fairly steep hill slopes are also common. Areas of steep slopes or low fertility are used for forestry and grazing. Even the severe damages resulting from modern warfare, which much of this region has experienced, are carefully repaired in the intervals of peace.

Dairying

In the cool summer lands to the north of the small-grain and livestock regions, in similar climates in the Great Lakes region and the Pacific Northwest of North America, and in parts of New Zealand and Australia, milk production is dominant. Crops and meat animals are usually of minor importance. Much of the farming land is devoted to hay and pasture, and since cultivation of row crops is minor, soil erosion is not usually a serious problem.

Dairying on a large scale depends on location near centers of population, which assures a market for the milk, butter, and cheese. In parts of northwestern Europe, notably in the Netherlands, marsh land, some of it below sea level, has been reclaimed and is used for dairy farming. This land presents special problems in conservation, in drainage, diking, and reclaiming of salt marshes.

In recent years the increase in the use of vegetable oils—cottonseed oil, peanut oil, and corn oil—as substitutes for dairy products has had an adverse effect on dairying. Faced with the loss of some of his market, the dairy farmer turns to the production of meat animals or to cash crops for which the climate and soil are suited.

Corn-belt farming

The corn-belt type of agriculture is similar to the small-grain and livestock type except that corn (or maize) is the dominant crop and the small grains and other crops play secondary roles. The principal area is in eastern United States, extending from, Nebraska to Pennsylvania, from the Great Lakes to the Ohio River. Other corn belt regions are to be found in southern Brazil, Argentina, the Po Plain of Italy, and the basins of Hungary and Romania.

The ideal corn climate has hot summers, including hot nights, and moderately heavy rainfall. A dry period before and during harvest time is an advantage. Corn is an excellent flattening feed for livestock and is used in a variety of forms for human food. The combination of corn, small grains, hay and pasture makes possible the large-scale production of beef, pork, lamb, and poultry. Most of the billions of bushels of corn produced are fed to livestock on the farm; only a small part enters world commerce.

The growth of the corn belts, the American one especially, was rapid and resulted in serious damage to the land in some areas. Much of the land was prairie before cultivation, but, in addition, large acreages of forest were cleared to make way for corn. In this hasty occupation of the land, even gentle slopes were eroded and fertility lost. Corn is a row crop and, unlike the, small grains, leaves the land nearly bare at harvest, subject to gulling or sheet wash. As a result, much of the land is poorer today than at the time of- settlement. Many cornbelt farmers do not think of themselves as permanent occupants of a certain farm.

A large proportion are tenant farmers who expect to move on in a few years. In most cases, however, erosion damage is not irreparable; careful conservation practices are halting excessive erosion and restoring a measure of fertility to the damaged lands. Many of the smoother areas have suffered little and some have had their fertility increased by careful methods of cultivation. The United States Corn Belt is probably the richest agricultural region in the world, in spite of the fact that some parts of it have been misused.

Wheat-belt farming

As agriculturalists moved into the drier lands, an unusual specialization developed. Because it became necessary to leave behind many of the crops which grew well.in the humid lands, the result has been a specialized wheat belt in which livestock, forage crops, and all the grains except one or two are not present. The great wheat farms with their use of heavy machinery have become an essential part of the agricultural system of the United States, Canada, Australia, and Argentina.

Many of these farms do not even have a house. The owner lives in town and simply attends to his acreage at the planting and harvesting seasons. Many farms have no livestock at all, although there is some grazing land available in the vicinity; on some farms it is the custom to graze animals on the stubble of the grain fields.

The extensive machine cultivation of the semiarid wheat farms greatly accelerates wind erosion and dust storms. Especially on the lighter soils, the pulverizing effect of the disc harrows and other cultivators causes severe losses of soil in times of severe drought and high winds. A single storm may remove all of the topsoil from many thousands of acres. And once the native grasses have been plowed under, it is very difficult to re-establish them.

Attempts to reduce wind erosion by the planting of shelter belts of trees have met with little success; rainfall is not sufficient for the growth of trees. In spite of government controls, there has been an increase in the acreage of wheat in marginal lands which should never have been cultivated. The average wheat farmer does not consider himself a permanent resident of the wheat belt; rather, he expects to make quick profits for a few years and retire to pleasanter climes. Wheat farming of this sort is sometimes referred to as "soil mining," since nothing is returned to the soil. Some aspects of wheat-farm conservation are unique. Wind erosion commonly leaves the land smooth, with little evidence of erosion that is apparent to the casual observer.

Dust storms from wheat farms reach cities, sometimes hundreds of miles away, to the distress of the city people and lead to demands for conservation measures by government agencies. The most effective low-cost method of reducing erosion in the wheat lands is strip cropping, with long, narrow strips of wheat alternating with untilled strips.

Plantations

The plantation provides the means of producing large quantities of cash crops which, with few exceptions, cannot be produced in a temperate climate. These crops include sugar cane, rubber, spices, coffee, cacao, and bananas.

The plantation depends upon cheap land, cheap, and some commercial crop which is in demand in areas of dense population in the temperate belt; coffee is a good example. For special crops certain conditions of soil, drainage, and climate are desirable. The plantation, then, is an adaptation of European agricultural practices to an entirely new situation. The plantations are so varied in different parts of the tropical and subtropical climates that an assessment of their effect on the land is particularly difficult.

In the United States Cotton Belt, which was once a plantation region, the system based on the cash crop has resulted in the most severe soil erosion in the country. Cotton cultivation, like corn farming, leaves the soil nearly bare after harvest, and the heavy rains attack even gentle slopes unless protective measures, such as terracing and strip cropping, are adopted. Nor does the substitution of the tenant "sharecropper" system for the plantation system improve matters.

2

Man and Technology

The history of the Old World is intimately bound up with the use of wheeled vehicles. The principle of the wheel was applied to transport probably a short time after its application to the making of pottery. As in the case of the potter's wheel, the first reliable evidence of wheeled transport comes from Sumer, in the Tigris-Euphrates Valley. The first wheeled vehicles were hearses and royal ceremonial chariots, both of which had solid wheels.

Made with a horizontal stock, the plows of the Near East and the Mediterranean area barely disturbed the soil's surface; but they were quite adequate in the light soils of these areas, where conservation of moisture (through dry farming) was of primary concern. Since the fields in these areas were cross-plowed, they were most often square in outline. Cultivation spread from the Mediterranean area to the heavier clay lands of northwestern Europe as a result of the demand for grain to supply the growing Roman Empire.

To cut and farm the sticky sod of northwestern Europe demanded a different sort of plow. From their experimentation in pre-Roman times, the northern Europeans had evolved a keel plow that was effective in sticky soils; when tilted in action by the plowman, it provided a turned sod. By the 11th century the mouldboard plow, which automatically overturns the sod rather than merely disturbing it, had come into use. (Although in common use for centuries, the mouldboard plow was not perfected until Thomas Jefferson designed its optimum shape.)

The fields of northwestern Europe became long and narrow as they were plowed by the keel plow and the mouldboard plow. Since cross-plowing was not necessary and since heavy soils were being worked, the farmers did not want to take their plowshares out of the ground frequently. As a result, there developed the typical field pattern of northwestern Europe, the strip-field. According to present knowledge, the first farming tool was a primitive sickle-made of a bone set with flint teeth-that was used by cave-dwelling farmers near Mount Carmel, Palestine, about 6000 B.C. Except for some technical improvements in shape and the application of its form to metal, this sickle remained the main harvesting tool until the invention of the mechanical reaper in the 1840's.

Pulleys and winches first appeared in Greece between the 6th century and the 4th century B.C. Both were applied in (and probably invented for) navigation and the theater, which used similar devices. It is not surprising that inventions should have arisen in connection with ships and stages, for sailing and the drama were of central importance in Greek life.

The mathematical bent of the Greeks may have led them to invent other machines by analogy. At any rate, the number of devices invented by the Greeks between 600 B.C. and 100 B.C. is a tribute to their imagination and skill. The Greeks distinguished five basic types of simple machines or devices for the conversion of motion: lever, wedge, winch, pulley, and screw.

Machines that provide motive power for other machines or tools are called prime movers. In describing the advance of technology, scholars distinguish five stages in the development of prime movers:

I Muscle power, mainly from human effort.

II Increased human and animal power.

III Power from the water mill and later from the windmill.

IV Power from the steam engine and internal-combustion engine.

V Atomic power.

The advance from one stage to another indicates the availability of more concentrated energy. The transition from stage I to stage II has sometimes been slow. For example, in ignorance of animal anatomy, the Romans attached the ox harness indiscriminately to horses and mules and obtained only four times the work of a man rather than the 15 times as much work gradually made possible after the 12th century by the use of a modern harness, independently evolved but very similar to the Chinese harness. By far the most important prime movers in the Middle Ages and the Renaissance period were wind and water mills. The experience gained in stage III, the age of mills, helped make possible the expanding settlement and the growing urbanization of northern Europe.

Water wheels apparently originated during the Ist century B.C. Used by the Greeks to grind grain (a constant task in all households), the first horizontal water wheels were limited to high-speed mountain streams, each wheel serving the needs of only one farmhouse. Although water wheels would not work in the sluggish rivers of Egypt and Mesopotamia, the use of the water wheel spread east to China and west to Ireland during the four centuries after its invention.

Not long after the discovery of the horizontal water mill, a Roman engineer invented a more efficient vertical mill. First using an undershot wheel placed in a stream, the developers of the vertical mill eventually found that the wheel did more work if the water overshot it and pulled it down. Aqueducts, common in areas influenced by Roman civilization, often provided the water necessary for the operation of vertical mills.

The introduction of the vertical mill into major urban centers is thought to have been prompted by a shortage of slave and pauper labour. By applying the principles of the water wheel in reverse, the paddle wheel was developed for moving boats and was first used in propelling warships. When the Goths besieged Rome in A.D. 537, they tried to starve the city by destroying the aqueducts that provided power for its grain mills. To counteract the effects of this destruction, the Romans put a paddle wheel between two boats anchored in the middle of the Tiber River and thus made the power of the flowing river available for the grinding of grain. Floating mills of this type still survive in the Po Valley.

Between the middle of the 10th century and the end of the 12th century, mills relying on water power became common in northwestern Europe. They were basic to the economic revival of the late Middle Ages and played a large role in the urbanization of northwestern Europe. The-following list of uses for water power from A.D. 1050 to 1380 is by no means exhaustive:

The earliest windmill-type objects were found by Chinese explorers in central Asia about A.D. 400. Consisting of scoops on wind-rotated prayer wheels, these early examples may have inspired the invention of the windmill designed for power purposes. The first working windmill for producing power for work was built by a Persian about A.D. 640; although 300 years later it was still confined chiefly to the Persian plateau, it spread to China in the 12th century. Windmills with four vertical sails attached to a horizontal shaft played a great part in the spread of rural settlement of the northern European plain. In areas far from dependable streams and in swampland, windmills could be used where water mills could not. First employed for grinding grain, windmills soon came to be used to lift water from poorly drained areas.

The first verified reference to a windmill in northern Europe dates from about A.D. 1180, and by the late 13th century windmills were common there. As the growing settlements in the gradually urbanizing portions of the North European Plain (Flanders to Paris and eastward to Lubeck) demanded fuel, peat bogs were exploited. As the peat bogs were dug, windmills were used (circa 1430) to pump water out of the shallow man-made lakes. Later, about 1600, windmills became common in the polder areas under reclamation by the Dutch.

The 15th century in northern Europe was a century of advance in the application of known power techniques to new uses, most of which uses had direct geographic significance. Because the growing amount of trade demanded better ports, new ways of dredging shallow existing ports were devised. New, methods of draining swamps were instituted, combining the Archimedean screw (a screw encased in a cylinder) and the windmill. Port facilities were improved by the addition of hoisting cranes (which ran on wooden rails) and swivel cranes, both of which offered greater ease of handling.

The internal network of canals in Europe for the most part dates from the 15th century and results from lock-gate systems which were much improved during that period. The successful development of various devices for draining and filling water locks permitted many new canals for inland navigation and many new connections between major navigable rivers. The combination of the Archimedean screw and the windmill that was introduced into the Low Countries in the 16th and 17th centuries was gradually replaced after 1640 by scoop wheels. In draining polders, scoop wheels worked in rows at a series of levels, lifting water from one level to a higher level until the water was pumped out of the polder.

The screw-windmill combination, however, continued to be used in the northeastern parts of the Netherlands, where small polders were very common. Of the larger polders under reclamation in the Netherlands between 1540 and 1690, about four fifths of it, or 650 square miles, was not drained by windmills and water lifts; it was reclaimed by man's diking off of the sea and his efforts in assisting accretion and accelerating sedimentation.

A second area of attempted reclamation in Europe included the Pontine marshes, an area of about 300 square miles to the southwest of Rome. Although reclamation efforts had been attempted, at the beginning of the 16th century the area was still a marsh with a notorious reputation as a malaria center. Between the beginning of the 16th century and the middle of the 17th century, there were four separate attempts to drain the Pontine marshes, each of which failed after the death of its chief engineer.

At the beginning of the 18th century, the plans of the last unsuccessful engineer were carried out by his son; but the plans failed, largely because of the opposition of the local inhabitants. In the Wash area of England some 700,000 acres, or 1,100 square miles, of low grassland and marsh began in the 17th century to be regarded as fit for reclamation. Since there was no pressure for increasing the amount of agricultural land, the reclamation was undertaken primarily as a financial venture. In 1630, with the advice of one Dutch engineer and reports by others who had once studied the project, the first drainage of the Wash area was attempted. The project, covering 307,000 acres, or about 480 square miles, was completed 23 years later in 1653. The area reclaimed was about 70 per cent of the total area reclaimed in Holland in the previous 150 years.

Reclamation of marshes in France followed the successful pattern developed by the Dutch both in their own country and, in the Wash of England. Because French reclamation was more politically centralized, the problems of organization and finance were solved speedily. Between 1607 and 1685 the marshes of Poitou, Normandy, Picardy, Languedoc, Provence, and other areas were drained and made into fine agricultural land.

Among the numerous inventions of the Industrial Revolution, most outstanding were the invention of the steam engine (the new prime mover mentioned by Forbes) and the discovery of the usefulness of coke in smelting iron. Between 1765 and 1788 James Watt of Scotland experimented with the principle of the steam engine, adding various refinements from time to time. In 1784 Henry Cort discovered that coke (coal partially burned in a shortage of air) could be substituted for charcoal in the process of smelting iron.

These two inventions had important geographical effects. The use of coke freed forge masters from their dependence on woodlands and permitted dozens of formerly scattered mills to be centralized around the four main coal fields of Britain. The use of steam as motive power allowed factories to be located away from high gradient streams and permitted their con concentration around the major coal fields. The economies that resulted from the coming together of iron mills and powered factories near the coal fields encouraged a growing supply of investment capital. Indus tries of all kinds expanded as profits were reinvested to take further advantage of the economies of centralization and the growing supply of cheap labour from the country side.

The "animating effects of cheap iron" were felt in shipbuilding and in agriculture, in weaving and in engineering. The geographical centralization of industries eventually led to central control or ownership of all the elements of the industrial process: from the mining of ore and coal to the processing of the final product of the mills. The cumulative effects of technology are well illustrated in the Industrial Revolution. Techniques used in one industry were often modified for use in a different industry; new machines and new processes often resulted from attempts to imitate old machines and processes.

The Industrial Revolution would hardly have succeeded had it not been for progress in the making of machine tools, or machine-making devices. Basic machine-tool types ive for gear makers, milling machines, planers, boy the rachines, lathes, and drills. Most of the fundamental thinking about machine tools, including even those used in modern times, was done between 1775 and 1850. The machine-tool industry was first developed by the British, but by 1850 the machine-tool industry in the United States was also advanced.

Many machine tools were made possible by the discovery of the cam principle (a cam is a wheel with a varying radius, which controls back and forth movements). Machines for making metal-boring screws for planing metal, and for making gears all utilize variants of the cam. After the introduction of the

Siemen's open-hearth process in 1856 and the coming into use of the Bessemer converter the same year, the price of steel dropped precipitously. The development of cheap steel vastly extended the range of machine tools, just as the introduction of cheap iron had previously extended the range of machines in general.

The growing urbanization caused by the Industrial Revolution was accompanied by better health and a reduced death rate. The English historian T. S. Ashton describes these improvements in his The Industrial Revolution 1760-1830:

> Many influences were operating to reduce the incidence of death. The introduction of root crops made it possible to feed more cattle in the winter months, and so to supply fresh meat throughout the year. The substitution of wheat for inferior cereals, and an increased consumption of vegetables, strengthened resistance to disease. Higher standards of personal cleanliness, associated with more soap and cheaper cotton underwear, lessened the dangers of infection. The use of brick in place of timber in the walls, and of slate or stone instead of thatch in the roofs of cottages reduced the number, of pests; and the removal of many noxious processes of manufacture from the homes of the workers brought greater domestic comfort. The larger towns were paved, drained, and supplied with running water; knowledge of medicine and surgery developed; hospitals and dispensaries increased; and more attention was paid to such things as the disposal of refuse and the proper burial of the dead.

The Industrial Revolution was accompanied by many changes in agriculture. The headlong migration of people to the cities was, of course, due in part to the pull of the cities and the opportunities offered by growing industries; but there was also a push from the villages as new and better methods of farming gradually came into use, making it possible for one man to do the work formerly done by three. New fertilizers were introduced, crop rotations became more complex, and there was a shift from smaller to larger properties through consolidation in most of England. Between 1760 and 1850 farm implements changed almost completely from homemade, relatively crude, and for the most part wooden hand tools to factory-made, standardized, and for the most part iron and steel-framed implements.

The development of farm machinery followed upon the enclosure movement, which resulted in the creation of large farms and estates in England; had the machinery been invented earlier, it would have been useless on the strips that characterized the three-field system. Self-sharpening, factory-made plows of cast iron that had been tempered in various improved ways were substituted for locally made plows. Scythes Were replaced by rather intricate machines for harvesting, and the flail disappeared in favour of a prototype of the modern threshing machine. Seed which was formerly broadcast on fields came to be sown in rows by seed drills.

The first reaper successfully adapted to farming was invented by Patrick Bell in 1826. Pushed by a team of horses, this reaper had sickle clips in front and a bar to bend grain over the clips (which features are like those still used on modern combines). This rather cumbersome reaper was not generally adopted, largely because Cyrus McCormick invented the pulled reaper in 1831.

Threshing machinery, using the age-old principle of the flail, had been invented by 1732, but the first successful threshing machine was built by Andrew Meikle in 1786. All the early threshers were stable machines to which the grain was brought. Many, threshers were driven by water power, while others were powered by horses walking around them. With the advent of mechanical reapers it was no longer necessary for great numbers of farmers and their families to cut the grain by hand; and the threshers eliminated the slow treading out of the grain by oxen. Reaping and threshing were reduced from a month-long task to one taking far less labour and requiring only two or three days to accomplish.

Threshing machines caught on very quickly and were widespread over the British Isles within a short time after their introduction. German agriculture lagged about a century behind that of Britain because German skill and technology were not actively applied to agricultural problems until the unification of Germany in the 1870's. Once a start had been made, however, German farmers rapidly became very scientific; they benefited from the knowledge of scientific farming that had been accumulated in Britain during the previous century, as well as from the ability of the German industrial setup to produce the materials and fertilizers needed by a modern agriculture.

As Britain industrialized, it was able to provide machinery for farming to areas which supplied wheat for its growing urban labour force. The interrelation between agricultural advancement in the United States and industrialization in Britain has been well pointed out by Sir Alexander Fleck: "In the 1870's the improvement of the marine steam-engine permitted an increase in the size of transatlantic grain cargoes, and this was decisive in bringing ruin to the British farmer. Conversion to an industrial society was thenceforth inevitable in Britain."

Among the many changes wrought by the increasing pace of technology in the past century, none has had greater effects upon population distribution than those permitting non-nomadic settlement of the world's grasslands. Similarly, no invention has had more radical effects upon modern life than the invention of the internal-combustion engine, which powers our automobiles and most of our trains. Without this engine, living in large cities would not be feasible.

Between 1850 and 1900 four major developments permitted a great rise in the population of Europe. In the first place, since industrial production grew more rapidly than population, the inhabitants of Europe were able to exchange their manufactured goods for food from abroad and for raw materials to maintain their industries. Also, although imported food could not have reached Europe in good condition before 1850, new methods of transportation, the introduction of steel ships, and the use of refrigeration to preserve food in them caused a revolution in food supply.

Thirdly, it was possible for Europeans to import large amounts of food after 1850 because there was an enormous increase in the amount of land farmed by Europeans living abroad. Much of this land was in the newly opened grasslands of Australia, Argentina, and North America, as well as in South Africa and the steppes of southern Siberia, and was for the most part actually farmed by European immigrants and sons of immigrants. In such other areas as Malaya, Ceylon, and West Africa, tropical products were made available from European owned and managed plantations.

In the intermediate areas or subtropical areas of Egypt and India, vast new irrigation projects permitted the growth of a number of industrial products, especially cotton, destined for the markets of western Europe. Finally, there was an increase in the productivity of European agriculture which was the result of better use of the land, increased use of machinery, new fertilizers, and improvements (through selective breeding) in the quality of farm animals and plants. Mechanization of the North American gasslands proceeded rapidly after 1850.

The McCormick reaper was selling at the rate of three per day; the wheat drill was in general use by the end of the decade; iron and steel plows were being sold in 1857 at the rate of 30 per day. However, it was not until a labour shortage was induced by the American Civil War that mechanization be came absolutely mandatory in the prairie areas of the United States.

In Australia the cultivation of wheat lagged behind that of the United States; because the soil was difficult to work, it was not until the stump-jump plow came into general use that the area devoted to wheat, was expanded. In both the Australian grasslands and the American prairies, wheat was grown on the humid fringe, while open-range cattle in great numbers grazed in the semiarid parts. To supply the growing cities with meat, railroads were extended into the semiarid cattle

It was the simultaneous development of iron ships and refrigeration which permitted the great expansion of meat growing in Australia. The iron ship dates back to 1787, when the barge Trial was built in England; in 1819 the first all-iron barge was completed, and the building of iron ships flourished from about 1850 to 1880. About 1880 the first steel ships marked the end of iron-ship construction: the last major iron ship, City of Rome, was constructed in 1881, while the launching of the warship Iris in 1877 is considered to be the beginning of the era of steel ships.

In 1851 the world's first refrigerating machine was set up for an Australian brewing firm. By 1873 refrigeration had been developed enough to allow frozen meat a full six months old to be eaten at an Australian public banquet, and in 1875 a slaughterhouse was constructed alongside ammonia-compression machinery for the freezing of meat. Thomas Mort, who had set up the first works for freezing meat in Sydney in 1861, told an audience in 1875: "Where the food is, the people are not, and where the people are, the food is, not. It is, however, within the power of man to adjust these things." Despite their skill and enthusiasm in developing refrigeration, the Australians were handicapped by being so far away from the European markets; with the limitations of the first refrigeration techniques, it was not certain that Australian meat would reach Europe in good condition.

The first successful shipment of chilled beef arrived in England from the United States in 1879. Two years prior to that, a successful shipment of frozen mutton had been made from Buenos Aires to France. Australian research in refrigeration eventually yielded results, making it possible fcr a vessel to arrive in England in 1879 with Australian beef which could not be distinguished from freshly killed English beef. By 1880, when New Zealanders could successfully send frozen lamb and mutton to London, the age of refrigerated ships was considered to have arrived.

The invention of the internal-combustion engine dates from Lenoir's engine of 1860. The Otto engine of 1878 was a slow-speed engine with horizontal cylinders. It became popular, and in a few years, more than 35,000 engines of this type had been sold all over the world. The modem automobile can be traced to two devices invented in 1886. Daimler in this year invented a vertical-cylinder, enclosed-crankcase engine with a carburetor. (In testing this engine on a bicycle, he created the prototype of the modern motorcycle.) In the same year Benz of Germany worked out his idea for an automobile.

The Benz automobile and the Daimler engine were combined a few years later into the first feasible automobile, the prototype of all automobiles made after that time. Eventually, rubber tires were developed for use on automobiles. Rubber had been known since the 18th century, when experiments were made with latex from the Amazon forest. In 1819 an Englishman named Hancock first became- interested in rubber, and by 1843 he had discovered a method of vulcanizing latex. His invention was stimulated by pieces of rubber he had obtained from Goodyear, the American inventor, and their patents were taken out within two months of each other in 1843 and 1844. Hancock developed most of the rubber goods which are used today, including shoes, rubber hose, valves, and rings. In 1888 an Irish inventor named Dunlop used the first air-filled tires on a bicycle for his 10-year-old son, and in 1895 the brothers Michelin used the first pneumatic automobile tires in the trial runs between Paris and Bordeaux. The use of air-filled tires on bicycles was so successful that the demand for rubber rose rapidly.

The rising price of rubber from the Amazon Valley stimulated the India Office of the British government to search for areas in southern Asia where rubber could be grown. In June of 1876 some seeds from rubber trees arrived at Kew Gardens from the Amazon Valley; by August 2,000 young rubber plants were on their way to Ceylon, to form the nucleus of the rubber industry in Ceylon and Malaya. The wisdom of this move became apparent when demand for rubber tires on automobiles skyrocketed some 25 years later, just as the rubber plantations in these areas were about to come into full-scale production.

The Industrial Revolution had been partly brought about by the invention of the steam engine in the 18th century, and the 19th century saw the invention of three more prime movers: the internal-combustion engine, the water turbine for generating electricity, and the steam turbine. The 20th century has produced still another prime mover—the gas turbine. These inventions have made possible an extremely flexible application of power to a wide variety of uses. For example, the increasing amount of power made available by thermal generation of electricity (through burning coal or petroleum to form steam) has had a great influence upon mining. Besides permitting greater flexibility in ore-recovery operations, the increased availability of power has made possible the processing of ores once considered worthless. Renewed life has come to many mining areas where tailings (waste ore) are reprocessed for low-concentrate ore once thought far beyond recovery.

The constant supply of power offered by the modern prime movers has permitted a wide diffusion of the mass-production system. The dockyards of Holland and Venice and the artillery schools of Spain and Austria in the early 18th century started the system of mass production; the system was subsequently improved when the idea of having completely interchangeable parts was developed in France by Le Blanc in 1785 and in Eli Whitney's gun factories in 1798. The American Civil War was the first conflict in which standardized factories utilized mass production for military purposes.

Replaceable parts for guns were made, and the manufacture of ammunition, canned food, and clothing on an assembly-line basis became common. The impetus of wartime conditions in the development of mass production should be noted. The increasing mobility of man is intimately related to the development of technology. The discovery that horses could be ridden and the discovery of the wheel; the invention of the chariot, the building of the Roman roads, the construction of seagoing vessels of large size, the development of canals in the late 18th century, and the subsequent development of railroads; the modern development of cars and various kinds of airplanes: all these mark a progression which has meant a constantly increasing physical mobility for mankind. As people develop the means of getting from one place to another faster or with greater ease, there are necessarily modifications of the cultural landscape; the changing cultural landscape in turn affects the evolution of technology. The student may find it profitable to attempt to calculate the future geographical effect of the various technological revolutions now under way.

The rising standard of living in the major industrial areas of the world, the development of nuclear energy as a dispersable source of energy, the possibility of transforming sea water into fresh water and using it to develop desert areas, the possibility of using sunlight as a direct source of energy by means of a solar converter or a solar stove, the decentralization of settlement and factories made possible by distilled and natural gas and by new methods of transmitting electricity more than twice as far as was formerly possible: add these together and imagine what changes they might induce in man's mobility and in the present worldwide distribution of population. It is evident from the above that any new source of energy contains within itself the possibility of a relocation of human settlement, particularly if it is a new fuel or a source of energy which can be substituted for a fuel.

Nuclear energy is the most spectacular of the modern prime movers, but its use is still hindered by the fact that it must first be converted into steam energy in a process which requires a large supply of water. Other sources of energy which have come into use in the last 80 years are electricity, industrial gas from coal, natural gas, and various petroleum products..

When iron was first produced, it did not have the effect of localizing industry to any great extent. Iron ore of various degrees of richness was widespread, and charcoal for smelting of iron could be made anywhere in the world that forests were found. However, agglomeration of iron smelters occurred, both in England and elsewhere, when coke was developed as a means of smelting iron and iron ore was consequently attracted to the coal areas.

The vast number of dispersed charcoal iron smelters gave way within a few decades in every country to the closely compacted agglomeration of steel mills. As a result of advances in iron technology, it is possible that production will become dispersed again. Since soft iron ore may be treated by oxygenation or some electrical method, the widespread availability of electricity is making possible the dispersal of new steel mills. Although coal and oil are often found in the same areas, the existence of one in a certain area does not guarantee the existence of the other in that area; in fact, many of the world's major oil fields are not near coal fields. Natural gas, which occurs in oil fields, may be used almost anywhere because it can be conveyed a long distance by pipelines.

As a result, it has not as yet caused any significant relocation of industry. With the development of electricity from hydroelectric sources, power has become even more widely available; most hydroelectric dams are located in areas other than those where coal is found, particularly in the Pacific Northwest. In addition, since coal in the future will probably be distilled rather than burned, industrial gas (which can be piped like natural gas) will be abundant in many areas. The geographical proximity of coal and iron will no longer be the main determinant in the location of industry.

In the past century, man has greatly increased his knowledge of the external world by developing or improving a variety of sensing devices which permit him to detect and record what is happening at a distance. The most familiar examples of distance-sensing devices are the telephone, radio, and television; records of what has been sensed by such devices may be kept on magnetic tape for future use and reference. Of great utility in enabling man to sense and understand his environment are a number of less common devices.

Recording barometers permit man to gather information on air pressure that is needed in making weather maps, and barometric devices have been used to record nuclear explosions in the atmosphere. Barometers sent aloft record the air pressure at various altitudes' and enable man to make pressure maps of, the upper air regions. Recording thermometers aid in the interpretation of pressure information. Seismographs (machines for recording earthquake waves) have permitted precise locating of earthquakes and occasionally have predicted when tsunamis (earthquake-generated sea waves) will arrive; the machines are also useful in studying the composition of the earth's core.

Radar performs a p, either oanalogous to the detection of earthquakes in e field of the . disturbances in the atmosphere andal welfar Both radar and seismographs may be used to gather information on nuclear explosions. The navigating of ships and aircraft and the guiding of missiles to strengthenea by very accurate inertial guidance voluntary efforts wo decades ago. Telescopes, both visu rces of voluntary types, have extended by many times the visibie universe.

The development of the transistor to replace bulky vacuum tubes has permitted the development of lightweight sensing and recording devices for exploration at great heights above the earth. The dangerously radioactive Van Allen belts were found and studied with such devices. Perhaps the most important development of all from the viewpoint of geographers has been the satellite-borne camera. The use of high-resolution film in Tiros photo satellites allows records to be made of the cloud cover over vast areas, which records may be instantly transmitted to ground stations for analysis.

The increasing accuracy of weather forecasting based on such information will permit far more efficient planning of man's activities on the earth. In the United States, geographers have been consulted on the planning of multi-spectral sensing of the earth's surface from orbiting spacecraft. In the early 1970's it will become technologically possible to obtain from a satellite a complete cattle census of Nigeria, for instance, or a rural housing census of Costa Rica, completely valid as of the last daylight orbit.

Paralleling the development of sensing devices used to gather information has been the growth of automation and automatic-control devices. The familiar household thermostat, one of the many automatic-control devices, uses the feedback principle that is common to them all. (Feedback control means that a machine, may work only between preset limits, and that there is constant communication between the control sensor and the machine-between the thermostat in our example and the furnace.) Automated factories now exist in many places, and large automated petroleum refineries can now be maintained "on stream" by half a dozen engineers. The problem of technological unemployment caused by automation is grave enough in the United States to have induced legislative action to provide funds to retrain displaced workers for other jobs.

Certain techniques and inventions of the modern world parallel those of the ancient world in their effect on man's life. The domestication of grains in the ancient world permitted nucleation of the population. With the domestication of grain, the people of the world were no longer exclusively hunters and gatherers; as agriculturalists, more people could live in less area. The experiments of modern scientists with the growth of algae may presage a similar change in the availability of food. The rise of cities in the ancient world was largely dependent upon the development of irrigation techniques for the growing of bread grains and rice. Early civilizations were almost by definition irrigation civilizations.

The mobility offered by the invention of the chariot in the ancient world permitted cities to combine into states and states to combine into empires. The high degree of mobility given the modern world by the automobile, the airplane, and other vehicles using internal-combustion engines has had similarly important effects. Finally, the development of the alphabet had an effect upon communications in the ancient world that may be compared to the influence that radio and television have had on the spread of information in the modern world.

All these inventions and techniques involve the extension of human effort. Irrigation in ancient times served to bring formerly unused areas into use, just as the modern irrigation of deserts serves to extend crop area. Likewise, the domestication of grains in the ancient world and the modern experiments with algae both involve an-extension of food varieties. Both the chariot of the ancient world and the internal-combustion engine of today involve an extension of physical mobility. And the alphabet, radio, and television all involve an extension of the mobility of ideas.

All these inventions utilize the inherent mobility and adaptability of man. Man can live almost anywhere on earth, can eat almost anything, and is apparently able to use almost any system of communication. The techniques and inventions discussed in this chapter do not represent ways of conquering or subduing nature (although Western man generally views them in this light); rather, they represent discovery of the possibilities inherent in the earth.

3
Search for Useful Minerals

At the beginning of the Stone Age, man discovered that stones picked up on the surface could be broken in such a manner as to produce sharp edges with which to cut, scrape, and dig. This discovery was the humble beginning of what has become a long and elaborate search for various kinds of useful minerals. Unlike the widespread soils and natural vegetation, minerals are likely to occur in small pockets, mostly beneath the earth's surface.

Since the beginning of industrialism, the search has been highly selective and well organized, making use of the latest scientific discoveries. Early man found a variety of minerals on the surface of the earth, some of which he could use with very little processing. Stones of various hardness could be fashioned into weapons and simple tools; salt was obtainable to season and preserve food.

Gold, silver, and copper occurred in the native metallic state; each could be pounded readily into any desired shape, but the fact that none of them could hold an edge very well led early man to seek other metals. Iron was available to early man only in the form of meteorites. Although most meteorites are stony, a few are metallic and are composed largely of iron and nickel, which man soon found to be much harder materials than gold, copper, and silver.

With the aid of fire, man gradually learned to smelt the ores containing metals which did not occur in the native state and thus to bring about great improvements in his tools and weapons. Some of the ores containing copper, tin, lead, and zinc could be reduced to metals very easily; moreover, since many of the ores were mixtures, the development of alloys was by no means difficult.

Modern industry depends upon an adequate supply of the right minerals. Every country or region with a well-rounded industrial development must possess or have success to essentially all of the fundamental minerals. These fundamental minerals may divided into groups according to their qualities and uses.

Iron and steel

Most modern industries use great quantities of power machinery, most of which is constructed of iron and steel. Iron, therefore, is the fundamental basic mineral, even though other minerals, such as aluminium, have been substituted to some extent for iron and steel for certain uses. The production of iron and steel calls for a special fuel, coke, most of which is derived -from high-grade bituminous coal. Good coking coal is not too abundant. There are large supplies of it in eastern United States, western Europe, U.S.S.R., and China, but many countries are almost completely lacking in it.

A very large part of the fuel used in manufacturing is consumed in the conversion of iron ore into pig iron and finally into steel. Petroleum and natural gas play a most important role in transportation and space heating all over the world; they are also basic raw materials for industry, especially for the chemical industries. The iron alloys, or ferroalloys, are not needed in such quantities as iron ore and fuels; however, with the modern demand for special quality steels, the iron alloys are of critical importance. Such iron alloys as manganese, nickel, tungsten, and vanadium, as well as some nonferrous metals, give steel the special qualities needed for certain purposes; a very hard steel may be needed for some uses and a very tough steel may be needed for others.

Nonferrous metals

The nonferrous metals include copper, lead, zinc, tin, aluminium, and magnesium. Copper is of great critical importance because of its use as a conductor of electricity. Lead has many uses, including the very significant one in modern times of shielding nuclear devices in order to protect workers from radioactive emanations. Aluminum and magnesium, often called light metals use of their low specific gravity, have produced in greatly increased amounts recent years. Aluminum, now produced large quantities, can, with suitable alloys, take the place of steel. Magnesium, somewhat lighter than aluminum but not quite.so stable, is produced in smaller quantities than aluminum from both ores and sea water.

Building materials

The construction minerals include building stone, sand and gravel, cement, gypsum, lime, asphalt, and asbestos. No one of these materials is especially critical since certain substitutions can be made. Cement can be produced wherever shale and limestone occur, either separately or in combination. Since building minerals are bulky and expensive to transport, it is essential to have sources of sand and gravel, cement, and other materials fairly close to important industrial areas.

Industrial chemicals

The chemical and industrial minerals are often discussed together. Examples of these minerals are the nitrate, phosphate, and potash minerals, all of which are used in the manufacture of fertilizers, Another significant mineral is sulphur, which is used in the manufacture of sulphuric acid. Sulphuric acid is necessary in the processing of mineral fertilizers. The mineral fertilizers make a great contribution to agricultural production in many parts of the world, and they also enter into industrial processes, including the production of chemicals and drugs. However, the geographical location of these minerals in relation to areas of industrial development is not so, important as the location of iron, fuels, and some of the ferroalloys and nonferrous metals.

Coal

Eight centuries ago the forests in parts of Europe and Asia were exhausted; much later some forests in the New World were in a similar condition. In the 13th century the forests of England had already been cut or burned to a very considerable extent, and fuel was very scarce. As a result, the English king, Henry III, gave his permission for the mining of coal, which practice had previously been carried on in Britain to a limited extent by the Romans.

Although this permission started the mining of coal on a rather large scale, there was still a good deal of opposition to the practice on the grounds that the burning of coal was unhealthful. Because of such opposition, Edwardson of Henry III, signed a decree prescribing the death penalty for anyone burning coal in London while Parliament was in session.

However, despite the belief that coal smoke was unhealthful, a belief confirmed by modem medical authority, coal mining and burning continued to increase, not only in England but in many other countries. There was little notice taken of coal in the Americas until 1665, when a French missionary made a map of the upper Mississippi Valley on which he marked the word coal.

Later, coal was discovered in a number of places in eastern North America, but it was generally considered to be of little value. Because of this low estimation of its worth, coal was once used in Pennsylvania as a road material, instead of gravel. During the 19th century, nearly 600 years after Henry III had given permission for the mining of coal in England, coal began to be used in eastern United States in some considerable measure. Just as in England, there was a struggle in the United States over the acceptance of coal as a fuel; then, once bituminous coal had been accepted, a struggle was necessary to get anthracite coal into use.

At the beginning of the 19th century coal was being imported from Europe the eastern United States in small quantities for the purpose of iron making, but there was little demand for coal for arty, other purpose.

A few years later coal war, mined and floated down the Susquehanna River to Lancaster County in Pennsylvania, where the people, never having seen coal, were skeptical about its use. Until a stove or grate was constructed to withstand the heat of the burning coal, there was little success with it as a domestic fuel; but once the grate and stove were introduced, the use of coal spread rapidly in the United States.

In the late 1700's and early 1800's, the invention and spread of the steam engine as a source of power had little effect upon use of coal; wood was still too abundant. Actually, coal did not reach its peak of production until the advent of electric power and the generation of electric power from' fuel plants.

Today, in many countries, petroleum has replaced coal as the chief source of energy. Few coal-burning locomotives are left in the United States and many power plants which formerly used coal have converted to oil. The distribution of good quality coal beds that could be readily mined has played an important part in industrial location.Coal attracts other minerals; iron ore goes to coal to be smelted and, as a result, the great iron and steel centers of the world have grown up in the coal fields: Pittsburgh, Birmingham, the Ruhr, the Donbas in the U.S.S.R., and Fushun, Manchuria, to mention a few. Other factors were involved, including the location of earlier metal industries based on wood as a fuel.

Petroleum

During the early part of the 19th century, especially during the 1830's, oil wells began to appear in the United States. These first oil wells were tapped accidentally during the process of drilling for water and brine. When oil was first discovered, there was little demand for it. Some oil was used for liniment and some was occasionally used in he crude state for fuel, but the demand was too small to encourage the further extension of well drilling. Because many advanced industrialized areas of the world did not have petroleum in any quantity, the large-scale use of petroleum as a source of power came somewhat later than the use of coal. Western Europe has never produced large quantities of petroleum and even small quantities were obtained only through deep drilling.

Many oil wells were drilled in different parts of the world before the 1850's, and petroleum was obtained from seepages in springs. In 1859 oil was discovered in the famous Drake well near Titusville in western Pennsylvania. The news of this discovery set fire to the imaginations of men concerning the great possibilities of oil, and as soon as the Drake well was in production, other wells were drilled in the surrounding country and oil production began on a commercial scale. The discovery of the well in Titusville started an eight-year oil boom; at first, production soared upward, but then it dropped to a mere trickle as the shallow wells were depleted. It was a period of great violence and lawlessness during which rival organizations endeavoured to destroy each other's pipelines and oil tanks. One of the first large uses of petroleum was for lighting the lamps of the world.

Prior to the 1860's, whale oil had been the most important source of light, and whale oil production in 1859 was 200 times as great as the production of oil. However, many people began to see the advantages of fuel from petroleum, which was both cheap and able to be distributed readily to different parts of the world.

With the invention of the internal combustion engine, the search for new oil fields became worldwide, leading to many discoveries in the United States, Mexico, Venezuela, Peru, the U.S.S.R., the Middle East, and Indonesia.

Exploration and drilling continue to expand, especially in offshore areas, with constantly improved techniques and although the total world consumption is more than 11 billion barrels per year, the estimated reserves, about 450 billion barrels in 1965, continue to increase. New discoveries of natural gas are being made also, and petroleum and natural gas together have become the most important source of power in many countries. By 1870 coal and petroleum had many small-scale uses, but they were still very primitively developed as sources of power.

Railroads were in use, but they burned wood; steamships had been developed, but many of them still us sails for auxiliary power. However, wood was becoming scarce near the cities, and more and more coal mines and oil wells were being opened up; man had, for the first time in history, the means of large-scale power development and industrial enterprises.

Mineral fuels

Today it would be difficult to overestimate the significance of the mineral fuels. More than half of the power in the United States comes from petroleum; coal is a close second in power production; hydroelectric energy, although increasing in importance, ranks a poor third. Still in an early stage of development, the large-scale use of nuclear energy is dependent on increased mineral production.

Many industrial operations require mineral fuels for which there are no substitutes. For instance, although it is possible to supply heat from sources other than coal, coal is essential to the smelting of any minerals and is an important source of many chemicals. Likewise, it would be very difficult, though not impossible, to run thermal-combustion engines in airplanes d automobiles without petroleum products.

What is the future of the world production of mineral fuels? So far as coal is concerned, there are large reserves which are usable, if not of the highest quality. Anthracite is scarce and high-grade bituminous is not too plentiful, but low-grade bituminous coals are present in enormous quantities and lignite, the lowest form of coal, is also available in large quantities.

On the other hand, although the best supplies of petroleum are relatively near exhaustion, new fields are being discovered almost every day. Reserves of petroleum are increasing in spite of the large production.

Although we are taking out a larger and larger quantity of petroleum from the earth every year, the estimated supply continues to increase because of the discovery of more liquid fuel. The question of how long this increase will go on has no certain answer, and it is very difficult to make an adequate statement about the future of world production.

We have to assume that the present estimates (which indicate an adequate supply for the next 40 years) are reasonably correct and that the demand for petroleum will keep on increasing from year to year. It is significant, of course, that in recent years the production in the United States has not been sufficient to supply the country's own needs. Sometime in the future, the world may very well utilize some form of nuclear energy, instead of petroleum and coal, as its main source of power. It is impossible to determine at the present time just exactly when this change will come and what it will mean.

The rank of a mineral in terms of its monetary value does not really express its total significance for modern industry and for the wellbeing of man. Such minerals as copper, lead, zinc, and iron are used in large quantities, whereas only small amounts of other minerals are needed. For example, germanium is a critical metal in the manufacture of transistors, which have replaced radio tubes in many instruments. Because it is a scarce metal, it is fortunate that only a very small fraction of an ounce of germanium is needed for each transistor.

The ferroalloys are of great importance. For instance, varying amounts of manganese, molybdenum, chromium, vanadium, and several other ferroalloys; are used in the manufacture of steel and are essential to its quality. The importance of a mineral then is to be judged not only by the total monetary value of the quantity used but also by its scarcity and the significance of its use.

A mineral which a country itself cannot supply in quantities sufficient to satisfy demand within the country is usually considered to be a strategic mineral in at country in the United States minerals considered to be strategic include uranium, tin, various steel alloys, manganese, industrial diamonds, and quartz.

The fact that many of the most useful minerals are relatively scarce in many parts of the world has a profound effect upon the political and industrial life of many nations. Faced with problems related to strategic minerals, the United States has taken two lines of action. In the first place, government subsidies have been directed toward the development of the low-grade deposits of such minerals as chromium.

Although chromium is a very important member of the ferroalloy group and chrome steel is a necessity for modern industrial development, the United States has no high-quality or rich chromium deposits of any size and its miners cannot carry on chrome ore production at the world price.

In order to encourage chromium production in the United States, the Federal government undertook a system of purchasing chrome ore from miners at a price well above the world market price. This purchasing program led to the development of many low-grade.

The rank of a mineral in terms of its monetary value does not really express its total significance for modern industry and for the wellbeing of man. Such minerals as copper, lead, zinc, and iron are used in large quantities, whereas only small amounts of other minerals are needed. For example, germanium is a critical metal in the manufacture of transistors, which have replaced radio tubes in many instruments. Because it is a scarce metal, it is fortunate that only a very small fraction of an ounce of germanium is needed for each transistor.

The ferroalloys are of great importance. For instance, varying amounts of manganese, molybdenum, chromium, vanadium, and several other ferroalloys are used in the manufacture of steel and are essential to its quality. The importance of a mineral then is to be judged not only by the total monetary value of the quantity used but also by its scarcity and the significance of its use. A mineral which a country itself cannot supply in quantities sufficient to satisfy demand within the country is usually considered to be a strategic mineral in at, country. In the United States minerals considered to be strategic include uranium, tin, various steel alloys, manganese, industrial diamonds, and quartz.

The fact that many of the most useful minerals are relatively scarce in many parts of the world has a profound effect upon the political and industrial life of many nations. Faced with problems related to strategic minerals, the United States has taken two lines of action. In the first place, government subsidies have been directed toward the development of the low-grade deposits of such minerals as chromium.

Although chromium is a very important member of the ferroalloy group and chrome steel is a necessity for modern industrial development, the United States has no high-quality or rich chromium deposits of any size and its miners cannot carry on chrome ore production at the world price. In order to encourage chromium production in the United States, the Federal government undertook a system of purchasing chrome ore from miners at a price well above the world market price.

The scarcity of strategic minerals has also been dealt with through extensive peacetime purchasing. In times of peace, when the strategic minerals are available on the world market, large quantities of the minerals may be purchased road and stockpiled as reserves to be used in case foreign supplies should ever be cut off. Such purchases are feasible in connection with minerals which can be stored without deteriorating through oxidation or contamination. One of the effects of scarcity of minerals has been greatly accelerated exploration of most parts of the earth.

Another effect has been the necessity for the "have" countries 'to aid the "have-not" countries with loans and grants to provide for mineral exploration and construction of processing plants. Still another profound effect has been the realization that no country is self-sufficient in all the essential minerals—a fact which has far-reaching political meaning.

Uranium

Many times in the past history of mineral Utilization has the discovery of a new use for a mineral revolutionized industry. For example, the whole fabric of energy production has been changed by the discovery made in 1945 that uranium, hitherto considered to be of limited use, was a source of energy a million times more concentrated than any material used before.

Of course, the impact of this discovery on world politics and warfare has been fully as important as its impact on energy production—perhaps even more important. It is possible now to provide a long-lasting source of energy within a very small space that in the long run may be cheaper than any other form of energy now in use. Nuclear energy is not an unmixed blessing, however, because its use produces harmful by-products and wastes which are very difficult to dispose of without contaminating air, water, and soil.

One current method of disposal involves filtering nuclear waste products and storing them in large iron tanks which are to he buried in the ground. Since these tanks are subject to rust, they eventually free the wastes, allowing them to contaminate the ground water. This method of disposal is also very expensive and results in such a high price for nuclear energy that the older sources of power-such as hydroelectricity and fuel-can easily compete.

If and when successful methods of waste disposal are developed at low cost, they will undoubtedly have a very strong effect upon the location of industry and population concentrations. It is possible that disposal may be worked out in connection with underground beds of rock salt. If such should be the case, the great cities of the future in the United States might be located in western New York, southern Michigan, and eastern Kansas.

Uranium is a fairly abundant metal in the earth's crust, but it occurs in such small concentrations that it is somewhat difficult and expensive to process. Discovered in 1789, uranium was considered a rare metal until recent worldwide exploration, on a scale never before known in history, resulted in the uncovering of many sources of uranium. Some of these sources represent very low-grade deposits which cannot be mined favourably at the present time but constitute possible sources for future use.

At the present time there are major uraniummining operations in the Congo, Canada, the United States, Australia, India, Portugal, and the U.S.S.R. In the Congo and in Canada (especially in the Sudbury and Great Bear Lakes areas) uranium deposits contain from 1 to 4 per cent uranium. Elsewhere, the concentration is much lower, but deposits containing less than 1 per cent have been successfully exploited in South Africa and in the United States.

In the future probably the most important use for uranium and the associated radioactive metals will be in the conversion of nuclear power into electrical energy. This conversion is very important for parts of the -earth which do not have good supplies of fuel or favourable conditions for hydroelectricity. However, for most countries the development of nuclear energy or atomic energy is not immediately critical since reserves of coal and petroleum are still fairly large and since hydroelectric energy, once developed, is continuous. The nuclear-power era which is being ushered in at the present time will have abundant source material in uranium, thorium, and other metals. The efficient use and economic utilization of these nuclear fuels is a problem which is currently being studied.

The amount of energy resource in nuclear fuels is much greater than the total amount contained in all the other fuels and sources of energy found in the earth itself. (Solar energy is not included in this comparison.) According to present developments, uranium will probably be the most important nuclear fuel for the near future. It is evident that the greatest changes in' the production and use of minerals in the near future will be in the field of the fuels, including the nuclear fuels. Petroleum, for example, will probably reach a maximum peak of production in the United States between 1965 and 1975 and in the world as a whole perhaps a few years later.

Thereafter, because of the increased demand per capita in other countries, it will be increasingly difficult for the United States to import cheap petroleum. The peak of world coal production will probably be around 1975 or 1980. After that, although there are large scale of coal in the United States and elsewhere, coal will be more expensive to Wine and transport to the places where it is used.

The rapid development of nuclear energy may modify some of these predictions by taking over some of the uses of petroleum and coal. Many people believe that in a decade or less the use of nuclear 14 energy will be common throughout world industry. However, any consideration of the various sources of energy must be concerned i not only with the total energy available from the sources but also with cost of the energy unit by unit. Thus, although nuclear energy is coming into use rapidly, the cost of its production may be too great for many years to allow it to become a replacement for coal or petroleum.

Mineral exploration

In order to supply the world's industries with various minerals, a search for new deposits is constantly being made in all areas of the earth with the aid of the most modern scientific apparatus. Unlike other re resources, minerals are often very scattered, some deposits are hidden far beneath the surface of the earth. The method of search depends upon the nature of the ore deposits. A few minerals, including common salt, building stone, and sand and gravel, occur in great abundance very near on the earth's surface.

On the other hand the ores of metallic minerals occur for the most part at some depth, although they may be exposed in part at the surface or there may be surface indications of their existence. The search for minerals is complicated by the fact that mineral deposits must have a certain degree of concentration in order to be workable. For example, iron ore must be at least 35 or 40 per cent iron before it can be profitably mined at the resent time. In some cases lower concentrations can be worked. Copper ores can be worked with a copper content of only 2 or 3 per cent. The concentration of the mineral required for mining depends upon the difficulty of the mining and smelting and on the value of the product. Gold, silver, and platinum can be mined profitably in low concentrations. Iron, aluminium and manganese must have higher concentrations. Early man explored for minerals by examining the materials that appeared on the surface of the earth. If he found a strange, heavy, black piece of rock, he might throw it into the fire to see what would happen. If the rock proved to be of some value, he might dig in the vicinity, hoping to find other specimens of the same material.

In the last century millions of prospectors have roamed over a large part of the earth's surface looking for mineral deposits of various kinds. Some of these prospectors have used fairly good methods and good instruments. Others have gone out with limited knowledge and a great deal of hope and, in many cases, have drilled or dug into the earth's surface without any definite indication of valuable deposits; despite its unreliability, this wildcat method has sometimes been successful. Since so many remote parts of the world have been carefully prospected, the discovery of new deposits is becoming less likely and less easy.

Nevertheless, there is indication that a great many mineral deposits remain to be discovered; discoveries have recently added -to the reserves of certain minerals which were considered to be approaching imminent exhaustion.

Today, prospecting for minerals is a very complicated process involving many devices. Most familiar of the recently developed techniques is the use of the Geiger counter in prospecting for uranium and other radioactive materials.

A Geiger counter enables a prospector without much training to find indications of these materials. Prospectors may also use the magnetometer, which measures changes in the magnetic force of the earth. Sudden changes indicate a variation in rock material and the possibility of large deposits of some heavy metal, probably iron or some related mineral. Prospecting with a magnetometer can be carried on from an airplane; and thousands of square miles can be covered, at least in a preliminary sort of way, in a few hours. If hopeful indications are found, more detailed investigations can be made on the ground, including test drilling to locate and evaluate the ore body. New methods and techniques are being introduced for the exploration of mineral deposits both on land and sea. Infrared and other special films are being used for airborne exploration. Soil samples, plants, and the water in streams are being analysed to suggest areas for further exploration. Seismic methods, long used in prospecting for petroleum, are now adapted to the search for metals. Induced polarization, a method of sending currents through the ground, makes it possible to detect metallic deposits 5,000 feet deep.

Minerals from the sea

As the resources on the land are diminished, man is turning increasingly to the sea in his search for minerals. All the minerals known to man are found in the sea, some of them in great quantities. It is estimated that the sea contains 50 trillion tons of minerals, including large quantities of magnesium, bromine, boron, uranium, silver, and gold, either in sea water or the ocean floor; the problem is how to mine or extract these and other minerals economically.

Large quantities of salt, magnesium, and bromine are now being produced from sea water. At present about one fifth of the world's petroleum is obtained from the shallow margins of the sea and it is estimated that this amount will be doubled in a few years.

Techniques of drilling under water are constantly being improved. Natural gas is derived from wells in the North Sea and new areas of natural gas production are being developed in the Gulf of Mexico. New devices for exploration of the ocean floor are also being improved. The submersible, a type of submarine, now makes it possible to dive to 20,000 feet and to photograph and pick up objects from the ocean floor. Millions of square miles of ocean floor are covered with metallic nodules, containing a variety of metals—manganese, nickel, iron, cobalt, and lesser quantities of molybdenum, vanadium, titanium, and other alloy minerals. It may be possible in the near future to mine these nodules with long-tube suction dredges, similar to the ones now in use in shallow water.

4
Cultural Diversity

The end of the cold war has created a series of tentative attempts to define "a new world order". So far, the only certainty is that the international community has entered a period of tremendous global transition that, at least for the time being, has created more social problems than solutions.

The end of super-power rivalry, and the growing North/ South disparity in wealth and access to resources, coincide with an alarming increase in violence, poverty and unemployment, homelessness, displaced persons and the erosion of environmental stability. The world has also witnessed one of the most severe global economic recessions since the Great Depression of the 1930s. At the same time, previously isolated peoples are being brought together voluntarily and involuntarily by the increasing integration of markets, the emergence of new regional political alliances, and remarkable advances in telecommunications, biotechnology and transportation that have prompted unprecedented demographic shifts.

The resulting confluence of peoples and cultures is an increasingly global, multicultural world brimming with tension, confusion and conflict in the process of its adjustment to pluralism. There is an understandable urge to return to old conventions, traditional cultures, fundamental values, and the familiar, seemingly secure, sense of one's identity. Without a secure sense of identity amidst the turmoil of transition, people may resort to isolationism, ethnocentricism and intolerance.

This climate of change and acute vulnerability raises new challenges to our ongoing pursuit of universal human rights. How can human rights be reconciled with the clash of cultures that has come to characterize our time? Cultural background is one of the primary sources of identity. It is the source for a great deal of self-definition, expression, and sense of group belonging. As cultures interact and intermix, cultural identities change. This process can be enriching, but disorienting. The current insecurity of cultural identity reflects fundamental changes in how we define and express who we are today.

The situation sharpens a long-standing dilemma: How can universal human rights exist in a culturally diverse world? As the international community becomes increasingly integrated, how can cultural diversity and integrity be respected? Is a global culture inevitable? If so, is the world ready for it? How could a global culture emerge based on and guided by human dignity and tolerance? These are some of the issues, concerns and questions underlying the debate over universal human rights and cultural relativism.

Cultural relativism is the assertion that human values, far from being universal, vary a great deal according to different cultural perspectives. Some would apply this relativism to the promotion, protection, interpretation and application of human rights which could be interpreted differently within different cultural, ethnic and religious traditions. In other words, according to this view, human rights are culturally relative rather than universal.

Taken to its extreme, this relativism would pose a dangerous threat to the effectiveness of international law and the international system of human rights that has been painstakingly constructed over the decades. If cultural tradition alone governs State compliance with international standards, then widespread disregard, abuse and violation of human rights would be given legitimacy.

Accordingly, the promotion and protection of human rights perceived as culturally relative would only be subject to State discretion, rather than international legal imperative. By rejecting or disregarding their legal obligation to promote and protect universal human rights, States advocating cultural relativism could raise their own cultural norms and particularities above international law and standards.

Largely through the ongoing work of the United Nations, the universality of human rights has been clearly established and recognized in international law. Human rights are emphasized among the purposes of the United Nations as proclaimed in its Charter, which states that human rights are "for all without distinction". Human rights are the natural-born rights for every human being, universally. They are not privileges.

The Charter further commits the United Nations and all Member States to action promoting "universal respect for, and observance of, human rights and fundamental freedoms". As the cornerstone of the International Bill of Rights, the Universal Declaration of Human Rights affirms consensus on a universal standard of human rights. In the recent issue of A Global Agenda, Charles Norchi points out that the Universal Declaration "represents a broader consensus on human dignity than does any single culture or tradition".

Universal human rights are further established by the two international covenants on human rights (International Covenant on Economic, Social and Cultural

Rights, and International Covenant on Civil and Political Rights), and the other international standard-setting instruments which address numerous concerns, including genocide, slavery, torture, racial discrimination, discrimination against women, rights of the child, minorities and religious tolerance.

These achievements in human rights standard-setting span nearly five decades of work by the United Nations General Assembly and other parts of the United Nations system. As an assembly of nearly every State in the international community, the General Assembly is a uniquely representative body authorized to address and advance the protection and promotion of human rights. As such, it serves as an excellent indicator of international consensus on human rights.

This consensus is embodied in the language of the Universal Declaration itself. The universal nature of human rights is literally written into the title of the Universal Declaration of Human Rights. Its Preamble proclaims the Declaration as a "common standard of achievement for all peoples and all nations". This statement is echoed most recently in the Vienna Declaration and Programme of Action, which repeats the same language to reaffirm the status of the Universal Declaration as a "common standard" for everyone. Adopted in June '1993 by the United Nations World Conference on Human Rights in Austria, the Vienna Declaration continues to reinforce the universality of human rights, stating, "All human rights are universal, indivisible and interdependent and interrelated". This means that political, civil, cultural, economic and social human rights are to be seen in their entirety. One cannot pick and choose which rights to promote and protect. They are all of equal value and apply to everyone.

As if to settle the matter once and for all, the Vienna

Declaration states in its first paragraph that "the universal nature" of all human rights and fundamental freedoms is "beyond question". The unquestionable universality of human rights is presented in the context of the reaffirmation of the obligation of States to promote and protect human rights. The legal obligation is reaffirmed for all States to promote "universal respect for, and observance and protection of, all human rights and fundamental freedoms for all". It is clearly stated that the obligation of States is to promote universal respect for, and observance of, human rights. Not selective, not relative, but universal respect, observance and protection.

Furthermore, the obligation is established for all States, in accordance with the Charter of the United Nations and other instruments of human rights and international law. No State is exempt from this obligation. All Member States of the United Nations have a legal obligation to promote and protect human rights, regardless of particular cultural perspectives. Universal human rights protection and promotion are asserted in the Vienna Declaration as the "first responsibility" of all Governments.

Everyone is entitled to human rights without discrimination of any kind. The non-discrimination principle is a fundamental rule of international law. This means that human rights are for all human beings, regardless of "race, colour, sex, language, religion, political or other opinion, national or social origin, property, birth or other status". Non-discrimination protects individuals and groups against the denial and violation of their human rights. To deny human rights on the grounds of cultural distinction is discriminatory. Human rights are intended for everyone, in every culture.

Human rights are the birthright of every person. If a State dismisses universal human rights on the basis of cultural relativism, then rights would be denied to the persons living under that State's authority. The denial or abuse of human rights is wrong, regardless of the violator's culture.

Universal human rights do not impose one cultural standard, rather one legal standard of minimum protection necessary for human dignity. As a legal standard adopted through the United Nations, universal human rights represent the hard-won consensus of the international community, not the cultural imperialism of any particular region or set of traditions. Like most areas of international law, universal human rights are a modern achievement, new to all cultures. Human rights are neither representative of, nor oriented towards, one culture to the exclusion of others. Universal human rights reflect the dynamic, coordinated efforts of the international community to achieve and advance a common standard and international system of law to protect human dignity.

Out of this process, universal human rights emerge with sufficient flexibility to respect and protect cultural diversity and integrity. The flexibility of human rights to be relevant to diverse cultures is facilitated by the establishment of minimum standards and the incorporation of cultural rights.

The instruments establish minimum standards for economic, social, cultural, civil and political rights. Within this framework, States have maximum room for cultural variation without diluting or compromising the minimum standards of human rights established by law. These minimum standards are in fact quite high , requiring from the State a very high level of performance in the field of human rights.

The Vienna Declaration provides explicit consideration for culture in human rights promotion and protection, stating that "the significance of national and regional particularities and various historical, cultural and religious backgrounds must be borne in mind". This is deliberately acknowledged in the context of the duty of States to promote and protect human rights regardless of their cultural systems. While its importance is recognized, cultural consideration in no way diminishes States' human rights obligations. Most directly, human rights facilitate respect for and protection of cultural diversity and integrity, through the establishment of cultural rights embodied in instruments of human rights law. These include: the International Bill of Rights; the Convention on the Rights of the Child; the International Convention on the Elimination of All Forms of Racial Discrimination; the Declaration on Race and Racial Prejudice; the Declaration on the Elimination of All Forms of Intolerance and of Discrimination Based on Religion or Belief; the Declaration on the Principles of International Cultural Cooperation; the Declaration on the Rights of Persons Belonging to National, or Ethnic, Religious and Linguistic Minorities; the Declaration on the Right to Development; the International Convention on the Protection of the Rights of All Migrant Workers and Members of Their Families; and the ILO Convention No. 169 on the Rights of Indigenous and Tribal Peoples.

Human rights which relate to cultural diversity and integrity encompass a wide range of protections, including: the right to cultural participation; the right to enjoy the arts; conservation, development and diffusion of culture; protection of cultural heritage; freedom for creative activity; protection of persons belonging to ethnic, religious or linguistic minorities; freedom of assembly and association; the right to education; freedom of thought, conscience or religion; freedom of opinion and expression; and the principle of non-discrimination.

Every human being has the right to culture, including the right to enjoy and develop cultural life and identity. Cultural rights, however, are not unlimited. The right to culture is limited at the point at which it infringes on another human right. No right can be used at the expense or destruction of another, in accordance with international law.

This means that cultural rights cannot be invoked or interpreted in such a way as to justify any act leading to the denial or violation of other human rights and fundamental freedoms. As such, claiming cultural relativism as an excuse to violate or deny human rights is an abuse of the right to culture.

There are legitimate, substantive limitations on cultural practices, even on well-entrenched traditions. For example, no culture today can legitimately claim a right to practise slavery. Despite its practice in many cultures throughout history, slavery today cannot be considered legitimate, legal, or part of a cultural legacy entitled to protection in any way. To the contrary, all forms of slavery, including contemporary slavery-like practices, are a gross violation of human rights under international law.

Similarly, cultural rights do not justify torture, murder, genocide, discrimination on grounds of sex, race, language or religion, or violation of any of the other universal human rights and fundamental freedoms established in international law. Any attempts to justify such violations on the basis of culture have no validity under international law.

The argument of cultural relativism frequently includes or leads to the assertion that traditional culture is sufficient to protect human dignity, and therefore universal human rights are unnecessary. Furthermore, the argument continues, universal human rights can be intrusive and disruptive to traditional protection of human life, liberty and security.

When traditional culture does effectively provide such protection, then human rights by definition would be compatible, posing no threat to the traditional culture. As such, the traditional culture can absorb and apply human rights, and the governing State should be in a better position not only to ratify, but to effectively and fully implement, the international standards.

Traditional culture is not a substitute for human rights; it is a cultural context in which human rights must be established, integrated, promoted and protected. Human rights must be approached in a way that is meaningful and relevant in diverse cultural contexts.

Rather than limit human rights to suit a given culture, why not draw on traditional cultural values to reinforce the application and relevance of universal human rights? There is an increased need to emphasize the common, core values shared by all cultures: the value of life, social order and protection from arbitrary rule. These basic values are embodied in human rights.

Traditional cultures should be approached and recognized as partners to promote greater respect for and observance of human rights. Drawing on compatible practices and common values from traditional cultures would enhance and advance human rights promotion and protection. This approach not only encourages greater tolerance, mutual respect and understanding, but also fosters more effective international cooperation for human rights.

Greater understanding of the ways in which traditional cultures protect the well-being of their people would illuminate the common foundation of human dignity on which human rights promotion and protection stand. This insight would enable human rights advocacy to assert the cultural relevance, as well as the legal obligation, of universal human rights in 'diverse cultural contexts. Recognition and appreciation of particular cultural contexts would serve to facilitate, rather than reduce, human rights respect and observance.

Working in this way with particular cultures inherently recognizes cultural integrity and diversity, without compromising or diluting the unquestionably universal standard of human rights. Such an approach is essential to ensure that the future will be guided above all by human rights, non-discrimination, tolerance and cultural pluralism.

5
Refugees

In a world where the nation state assumes increasing importance, nationality is an essential attribute of a person's material and moral well-being. Normally every individual has a nationality, and one only. In reality, some people have several nationalities, while others - much more numerous - have none. They arc stateless. No State considers them as one of its own. There are hundreds of thousands of people in this predicament spread across the globe, although exact figures are difficult to determine. The subject of statelessness has received little detailed attention from international organizations.

Some people are born stateless, others become stateless in the course of their lifetime. Certain countries confer nationality according to the principle of birth within national boundaries (jus soli), others confer it according to filiation (jus sanguinis). With regard to statelessness after birth, there may be several causes. An individual may lose his or her nationality without acquiring a new one as a result of an extended stay abroad, or due to the ceding of territory, or for some perceived disloyalty, or as a punishment, or simply through marriage to someone of a different nationality. In this last case, it is usually women who lose their nationality. From a legal point of view, statelessness may be considered as the faulty expression of national legislations on the question of nationality and the principle by which the law of each State fixes the acquisition, attribution or loss of nationality.

Statelessness is not merely a legal problem. It is also a consequence of migrations, of movements of population, of forced exodus and of various forms of persecution. The frequency and growing magnitude of these problems in recent years explains, to a large extent, the renewed interest in statelessness, and particularly in de facto statelessness.

De facto stateless people, as distinct from de jure stateless people, do theoretically have a nationality. But they have lost all real links with their country of origin and no longer have its protection, either because the country in question refuses to afford assistance and protection, or because the people in question have renounced it. Their situation is similar to that of refugees who, while keeping their nationality, no longer benefit from the effective protection of their State. Stateless people are not, however, all refugees, and, conversely, refugees are not all stateless. In practice, the distinction between de jure and de facto statelessness is not always evident and this has sometimes led to the use of the terms 'de jure non-protected persons' and 'de facto non-protected persons', or 'technically stateless persons' and 'politically stateless people', in order to demonstrate the origin of non-protection.

As de facto stateless persons, there are also those who have never had a nationality. In particular, the case of many Palestinians serves as illustration. There are also nomads and gypsies who live on the fringes of the nation state and who, as it were, embrace statelessness voluntarily. Their situation raises particular problems which the Council of Europe, for example, is trying to solve. Its recommendation of 1983 calls on States to retain certain criteria, such as the country of origin of immediate family or of frequent periods of residence, in order to facilitate their membership of a State and to allow them to travel freely.

From a humanitarian point of view, statelessness, whether de jure or de facto, is a major preoccupation. The stateless are part of those unwanted people who are refused right of entry to countries or who are turned back at borders according to political or economic circumstances. If they are taken in, they often live in uncertainty for many years. Statelessness is frequently transmitted from one generation to another. Nationality is not only the right to a passport and material advantages. It also confers upon the individual an identity and the sense of belonging to a community - elements without which a person remains vulnerable and uprooted.

As a result of events during the inter-war years and the Second World War, the Universal Declaration of Human Rights proclaimed the right to a nationality as a fundamental human right. Article 15 states:

1. Everyone has the right to a nationality.

2. No-one shall be arbitrarily deprived of his nationality, nor denied the right to change his nationality.

But much effort is still required before this right is fully recognized. In recent decades, the international community became aware of the problem of stateless people at the same time as it was preoccupied with the fate of refugees. However, the two issues were dealt with separately by the United Nations. It is now becoming increasingly apparent that, in practice, the stateless are less protected than refugees.

The 1954 Convention of the United Nations relating

to the Status of Stateless Persons, formulated along the same lines as the 1951 Convention relating to the Status of Refugees, confers a special status on the stateless and guarantees them minimum standards of treatment. Stateless persons are thus granted the same rights as nationals in certain areas: the right of access to courts of law, industrial and intellectual property, education, assistance and social security. For other rights and activities, they should be granted treatment no less favourable than generally accorded to aliens as regards movable and immovable property, the right of association, the right to engage in wage earning or unsalaried employment, the right to practise a liberal profession, the right to housing, education up to and beyond primary level, and freedom of movement. The Convention also requests States to issue to stateless persons identity papers and travel documents and to take measures to facilitate their assimilation and naturalization.

The treatment of stateless persons and refugees has many similarities. There are, however, certain differences which make statelessness a less advantageous situation than that of refugees. In the event of illegal entry or presence, only refugees are exempt from penalties. They are also guaranteed freedom from expulsion or forcible return to a country where they risk being persecuted. However, the 1954 Convention offers no more than a statutory arrangement for stateless persons. It does not offer a definitive solution. It does no(look into the problem of nationality which is, after all, the only means to ensure true and lasting integration of the stateless into a national community.

In this sense, the 1961 Convention on the Reduction of Statelessness took a step forward by attacking the problem at its source. Stateless children should be granted the nationality of the State in whose territory they are born. This would avoid perpetuating statelessness from one generation to the next. If all countries were to conform to this principle, namely that of jus soli, there would be far fewer stateless persons. But this is hardly the case. Many countries apply the rule of jus sanguinis. They ensured that their own interests were preserved in the 1961 Convention. The co-existence of both principles weakens the efficacy of the Convention in relation to its avowed aim.

The Convention also seeks to avoid cases of statelessness which are the result of voluntary omission, of a change in civil status, of the stripping of nationality or of the transfer of territorial sovereignty. The Convention essentially ensures that no person shall lose his or her nationality without a guarantee of acquiring another. However, even here, the prerogatives of sovereignty have been assuaged, and exceptions introduced which largely reduce the scope of the principles in question. This is certainly the case of persons who may lose their nationality as a result of extended residence abroad. Others may be deprived of nationality if, for example, they represent 'a serious threat' to the 'essential interests' of the State. These notions are not spelt out in more detail, and their Interpretation often results in injustice and arbitrary measures.

De facto statelessness has been particularly overlooked in international conventions. Protection under the 1954 Convention is only aimed at people who have obtained status as stateless persons. Other than those who also have refugee status, and thereby enjoy more favourable conditions, the majority of de facto stateless persons are without protection. Even in war time when the number of refugees and stateless persons increases, the 1977 Geneva Protocols expressly qualify as protected persons only those stateless persons considered as such before the outbreak of hostilities, namely the de jure stateless.

A recommendation included in the Final Act of the Conference on the Status of Stateless Persons invites the Contracting States to extend the treatment accorded by the Convention to de facto stateless persons who have renounced the protection of their own State. Among the resolutions included in the Final Act of the Conference on the Reduction of Statelessness is one which recommends that persons who are stateless de facto should be treated as stateless dejure to enable them to acquire an effective nationality. But these are only recommendations which are in no way binding upon States.

Despite numerous compromises and concessions, the Conventions on Statelessness have not received the number of ratifications one might have hoped for: only 35 States have ratified the 1954 Convention and 13 that of 1961.

To the weakness of standard-setting, one must add negligence by existing institutions. Stateless persons are almost alone therefore when facing the State. They frequently leave their country without any resources or valid identity papers. They may either be forcibly returned, or admitted temporarily and then rejected and escorted to the frontier. They may be mistreated - without any international organization offering assistance. The Convention relating to the Status of Refugees III was able to benefit from the establishment of the Office of the United Nations High Commissioner for Refugees (UNHCR) which was entrusted with its supervision and application. UNHCR has the mandate to assist refugees, ensure their legal protection and follow up with the authorities of host countries. There are no such mechanisms for stateless persons.

The 1961 Convention did foresee the establishment within the framework of the United Nations of a body to which persons claiming the benefit of the Convention might apply. UNHCR was so designated by the UN General Assembly (Resolution 3274 (XXIX) of 10 December 1974 and Resolution 31/36 of 30 November 1976). But no mention is made of UNHCR's competence with regard to the application of the 1954 Convention and to measures of assistance towards those who have not yet been deemed stateless persons. Moreover, UNHCR, which could in practice go beyond conventional limits, has remained somewhat indifferent when it comes to the fate of the stateless. The term stateless person hardly ever appears in UNHCR publications - a fact which, together with the doctrinal vacuum in this particular area, only serves to heighten general indifference towards a problem which should, rather, inspire in human terms the same compassion as that shown to refugees.

The solutions provided so far by the international community with regard to the problem of statelessness have been found wanting and inefficient. They have been formulated more from the point of view of a State's prerogatives and sovereignty than of individual human rights. It is this imbalance which today requires correction through a new approach which takes into account the many dimensions of the problem. Action must be aimed at prevention and elimination.

With regard to nationality, the law needs to be strengthened and better defined. The 1961 Convention does not express an international consensus. Few States (only 35 of them) participated in the Conference and only 13 have ratified the Convention. The elaboration of a new instrument might allow the emergence of a consensus and a new standard which, in any case, would serve to influence national legislation.

The 1961 Convention was a compromise between those countries which, in turn, supported either jus soli or jus sanguinis. Ideally, any future international instrument should go beyond compromises to a comprehensive approach to the elimination of statelessness. It would be advisable, for example, to adopt a single criterion for the acquisition of nationality: that of jus soli. The 1969 American Convention on Human Rights, for example, recognizes the right of any person to acquire the nationality of the State in whose territory he or she was born.

The ability to deprive a citizen of nationality should be forbidden absolutely if it leads to statelessness. There are other measures that can be taken against those who go against national interests. A State should not turn away its own citizens, or ex-citizens who have no other nationality.

In the case of ceding of territory, States must ensure that the inhabitants do not become stateless. They might give the inhabitants, for example, the right to choose between the nationality of the successor or the predecessor State.

The best way to eliminate statelessness is to allow stateless persons to acquire a nationality. States should envisage advantageous ways of naturalization and distinguish them from those usually offered to foreigners. In the event of a country of residence not granting its nationality to a stateless person, the United Nations should look into ways and means to resolve the problem. It is up to the United Nations to initiate and encourage the international community to be more sensitive to the problem of statelessness.

Mechanisms to protect and assist stateless persons should be established urgently. It is possible, for example, to envisage, through a protocol to the 1954 Convention, the-creation of a High Commissioner for Stateless Persons, However, it would be more appropriate to confer this task upon UNHCR. This would be a way not only of gaining from its experience in the field of protection, but of saving time, money and energy. UNHCR already has been given by the UN General Assembly the functions foreseen in the 1961 Convention. It should also monitor the application of the 1954 Convention. The mandate of UNHCR is sufficiently flexible. In the same way that it is charged with the task of protecting refugees and displaced persons who do not fall under the definition of the 1951 Convention, it should be empowered to extend its activities to all stateless persons, de jure or de facto. It is important to recall in this connection that UNHCR has on-going activities not only in countries which are party to the Convention, but also in many which have not ratified it.

We believe it would be helpful if UNHCR could assume the following tasks.

— Promote the ratification of international Conventions relating to statelessness.

— Monitor the application of Conventions and report to the General Assembly on the state of application of Conventions and of its activities in favour of stateless persons.

— Collect all possible information concerning stateless persons, their number and whereabouts. The task should be all the easier, in that UNHCR already undertakes this for refugees and that within national administrations, it is often the same department which deals with refugees and stateless persons.

— Co-operate with governments with a view to improving the condition of stateless persons and reduce their number through the adoption of permanent solutions through appropriate intergovernmental agreements.

— Assist stateless persons in administrative matters and inform them of their rights.

— Raise public awareness of the problems of stateless persons and promote the reduction and elimination of statelessness.

— Include the needs of stateless persons in its public information and fund-raising activities.

The 1961 Convention foresees the jurisdiction of the Inter national Court of Justice in the event of disputes concerning the interpretation or application of the Convention. But the prerogative of seizing the Court is reserved for States. It would appear timely to set up an international authority to which stateless persons themselves could apply in order to uphold their rights. Nationality is not a simple affair of state. It is also a human right.

Statelessness is a source of political and socio-economic tension which affects the country of origin, the country of residence and neighbouring countries. Regional organizations should become more involved in the problem and contribute to its solution through initiatives adapted to local conditions. Regional mechanisms to control human rights could also be used to ensure respect for the right of every individual to a nationality and related rights such as that of choosing one's place of residence, of freely entering one's own country, or of not being expelled from it.

According to the Statute of the Office of the United

Nations High Commissioner for Refugees (UNHCR), a refugee is a person who, owing to a well-founded fear of persecution for reason of race, religion, nationality, membership of a particular social group or political opinion, is outside the country of his nationality and is unable or, owing to such fear, unwilling to avail himself of the protection of that country.

The essential problems of refugees, and the necessary solutions to them, have been the same throughout history and around the world: refugees need to be fed, clothed and sheltered; the want to go back home if they can, and often need help to do so; and they need to find new homes if they cannot return to their old ones. The Office of the United Nations High Commissioner for Refugees was established by the General Assembly in 1950 to protect refugees and promote durable solutions to their problems. UNHCR depends entirely on voluntary contributions from Governments and private sources for its programmes seeking to assist as many as possible of the more than 12 million refugees in the world.

The basic function of UNHCR is to extend international protection to refugees who, by definition, do not enjoy the protection of their former home country. UNHCR seeks to ensure that refugees receive asylum and are granted a favourable legal status in their asylum country. An essential element of this legal status is the safeguard provided by the generally accepted principle of non-refoulement, which prohibits expulsion or forcible return of a person to a country where he may have reason to fear persecution. UNHCR also seeks to ensure that persons claiming to be refugees are identified as such so as to enable them to take advantage of the standards in international instruments or under international law.

The legal status of refugees has been defined more specifically in two international instruments-the 1951 Convention relating to the Status of Refugees and its 1967 Protocol which define the rights and duties of refugees and contain provisions dealing with such matters as protection from refoulement, unlawful expulsions and detentions, and the refugee's employment, public education and assistance, artistic rights and industrial property. In regard to many of these matters, refugees are to receive the same treatment as nationals of their country of residence.

Material assistance is also essential, to enable refugees or displaced persons to achieve permanent solutions to their problems, whether it be through voluntary repatriation, local settlement in the country of first asylum or migration to another country. Measures vary widely according to need. Integration through rural settlement is in general the best solution for refugees of rural background who cannot hope to repatriate for some time; in Africa, in particular, UNHCR helps refugees to settle on the land through the establishment of new rural communities or through strengthening the infrastructure of areas where the refugees can be absorbed spontaneously into the local population.

Assistance in voluntary repatriation may involve a large-scale transportation and settlement operation, or may simply entail the payment of travel expenses for individuals.

Resettlement of Latin American and Indo-Chinese refugees are examples of instances where UNHCR's role has been to negotiate with Governments to encourage the admission of refugees or displaced persons into their territories. UNHCR has for many years helped to organize special resettlement schemes for refugees who could not qualify under normal immigration criteria because of physical or social handicaps. In addition, family reunion is an essential activity. Finally, counselling plays an important role in helping individual refugees find the most appropriate solutions to their problems.

UNHCR was awarded the Nobel Peace Prize in 1954 and in 1981. The Second International Conference for Assistance to Refugees in Africa (ICARA II) was held at Geneva in July 1984. It adopted a Declaration and Programme of Action aimed at initiating a long-term strategy to deal with Africa's 4 million refugees and returnees. Going beyond the first International Conference (ICARA 1) (Geneva, 1981), which focused primarily on emergency assistance, ICARA 11 linked humanitarian aid to refugees with the need to help host countries develop their social and economic infrastructures.

At a spring 1989 meeting, the U.S. Ambassador to Hungary, Mark Palmer, told a group of journalists to keep an eye on Hungary's treatment of refugees. Budapest, he said, .'had just signed onto several U.N. treaties covering refugees' rights, obligating Hungary to protect all émigrés fleeing persecution. Palmer predicted, East Germans would soon begin slipping across the border to seek shelter. If Hungary chose to live up to its obligations, a major Soviet-bloc row might ensue.

Summer and autumn newspaper reports and television broadcasts showed busloads and trainloads of ecstatic East Germans fleeing their repressive homeland for the democratic West. It is hard to remember a year in which refugees themselves helped deliver such a final blow to a totalitarian regime; it is hard to recall a time when the entire Western world drew such inspiration from those fleeing communism.

But like all else over the past year, the excitement and promise of events in Europe threaten to draw attention and resources away from developing world problems. While tens of thousands of Eastern Europeans have found homes in Germany or elsewhere, an estimated 15 million refugees linger in desperation in camps around the world-6 million more than only six years ago [National Geographic News Service, 1/17/90]. The United Nations estimates another 15 million people may be internally displaced within their own countries by hunger or civil strife, many beyond the reach of international aid [A/44/520].

As numbers have risen, shelter and aid have not kept pace. The United Nations High Commissioner for Refugees' (UNHCR) budget has grown only 25 percent over the past five years-half of that needed to care for the new migrants under the agency's protection [The New York Times, 3/18/901. The U.S. share of contributions has fallen over the past five years from 27 to 22 percent [ibid.]. The agency estimates its 1990 needs at $700 million but so far expects no more than $550 million from donor nations [ibid.].

One problem facing the U.N. and other efforts to protect refugees has been the narrow interpretation of who constitutes a refugee, as defined by U.N. and other treaties. The Universal Declaration of Human Rights, considered binding on all states, "affirms a full range of civil, political, economic and social rights for all persons, citizens and refugees alike. It also significantly provides the right to seek and enjoy asylum, the right to freedom of movement, and the right to a nationality" [Refugee Policy Group, 1/90]. Other pacts, such as, the International Covenant on Civil and Political Rights, cover all people within a nation's territory and prohibit expulsion of aliens or limitations on movement.

But other pacts that spell out refugee rights define protected refugees more narrowly. The 1950 Statute of the UNHCR and the 1951 Convention Relating to the Status of Refugees prohibit forced return of refugees and state they must be treated almost as well as Citizens. Refugees must be accorded free movement, ownership of property, legal rights, access to education, work, and other human rights. The instruments define refugees as people with a well-founded fear of persecution based on racial, religious, national, social, or political grounds. This definition covers some-but not all-émigrés fleeing civil strife, war, or poverty related to political persecution or ethnic conflict.

The General Assembly adopted the Declaration on Territorial Asylum in 1967. It seeks to bar states from rejecting asylum seekers at the nation's frontiers and would bar the return of asylum seekers to countries in which they would be persecuted. But like all other Assembly resolutions, it is nonbinding. Efforts to draft a binding convention collapsed in 1977. Many of the problems facing refugees stem from disagreement over these definitions of who deserves asylum. Politically it has been a good year for tens of thousands of anticommunist, generally educated Warsaw Pact refugees whose flight has vindicated years of Western criticism of Moscow and its satellites. But it has been a terrible year for more simple masses fleeing persecution and war, or merely life-threatening poverty, in the developing world. Countries continue to show a preference for refugees fleeing the host nation's enemies as well as for migrants with good skills and education. Welcoming countries also tend to favor exiles supported by powerful, domestic groups. How to define a refugee-as opposed to someone merely seeking economic opportunity—continues to be an issue, one that some governments use as a fig leaf to cover their choice of which refugees to admit and which to reject. On December 12, 1989, the British government began a secret operation to repatriate forcibly up to 44,000 of the 55,000 Vietnamese exiles who have flooded Hong Kong over the past few years. A document drafted by the region's first-asylum countries for an International Conference on Indo-Chinese Refugees earlier in 1989 had established a process for screening Indochinese boat people to distinguish political refugees from economic migrants, but the United States and Vietnam rejected the notion of forced repatriation for those denied refugee status. In the summer of 1990, as the refugee/migrant waves continued and Washington and Hanoi stood their ground vis-a-vis forced repatriation, the countries of first asylum threatened to refuse landing rights to all new arrivals.

U.S. refugee policies-intended to protect those in political danger reveal similar contradictions. Washington usually accepts those fleeing governments it opposes or refugees who bring skills and education with them. Of 125,000 annual openings for refugees, the lion's share goes to Soviet-bloc and other exiles fleeing communism. The United States rejects 97 percent of asylum requests from friendly El Salvador, though 60,000 civilians have been killed there in the past decade. Only six of 20,000 Haitian boat people intercepted through October 1989 were admitted. The United States accepts 1,000 refugees from Marxist Ethiopia, and a handful more from the rest of the continent, though Africa houses nearly one-third of the world's refugees.

Other examples of ambiguous refugee policies abound. There are 320,000 Cambodian refugees in Thailand. Most live as virtual prisoners with no freedom of movement in camps along the Thai-Cambodian border, deprived of rights accorded them under various refugee pacts. UNHCR has little direct access to these camps; and many Cambodians have been pressed into military service by the brutal Khmer Rouge faction, which runs its camps like fiefdoms.

In late 1989, Arab countries mounted a major diplomatic effort to convince the Soviet Union and transit countries not to aid the flow of Soviet Jews to Israel, claiming they would colonize the occupied Arab territories-this despite the fact that only a few hundred so far have chosen to settle in the occupied lands.

Most countries have done little to aid tens of thousands of Kurds fleeing intense persecution and chemical bombardment in Iraq for fear of alienating the influential Baghdad regime or Turkey, a member of the North Atlantic Treaty Organization.

Eritrean migrants in Gulf states are often carted out to sea and dumped on islands off the Ethiopian coast, where conditions often lead to death [The Economist, 12/23/891. Both Sudan and Ethiopia use food and aid as weapons against their own and refugee populations in their countries' civil wars.

The United Nations itself must negotiate these tricky political issues supporting refugee rights without aggravating major donors or countries sheltering exiles. Critics say it sometimes plays the political game too well. The High Commissioner for Refugees, Jean-Pierre Hocké, resigned on October 26, 1989, amid widespread charges that he undercut the agency's management and had misused hundreds of thousands of dollars for entertainment and travel [The Washington Post, 10/27/89]. But Hocké, a Swiss native, was also accused of catering to the interests of UNHCR's big donors at the expense of the interests of refugees.

In January 1988 he had ordered the burning of more than 130,000 copies of a UNHCR magazine containing an article critical of West Germany's refugee policies. Roger

Winter, head of the U.S. Committee for Refugees, also charged that Hocké "took the low road on the function of protecting refugees," refusing to criticize governments when they deported controversial exiles. Winter cited U.S. actions to keep Haitians out and Thai and Hong Kong policies of evicting Indochinese refugees, issues on which he said Hocké was "conspicuously silent."

The new High Commissioner, Thorvald Stoltenberg, has been a more vocal advocate of greater support for those under his protection. Complaining of budget constraints, he told The New York Times on March 18, "1 can't fulfill my mandate-levels of care are falling, I have only half the resources per refugee today that we had in 1980."

The General Assembly, for its part, continues to pass dozens of resolutions each year calling attention to specific refugee issues and the needs of refugee groups around the world. On December 8, 1989, it adopted 11 resolutions on Palestinian refugees who fall under the mandate of the United Nations Relief and Works Agency (UNRWA). The resolutions appealed for more funding for UNRWA, demanded that Israel not resettle any refugees in the Occupied Territories, affirmed the need for scholarships and strengthened educational facilities, and in general sought to further the Palestinians' cause. While the resolutions may have been worthy, they reflect the fact that only refugees with vocal political supporters-in this case the Arab group-tend to get major political support in the General Assembly.

On December 15 the Assembly adopted by consensus a resolution supporting UNHCR. It called for all states to grant asylum in accordance with international law, noted the fact that many of those seeking asylum are held in detention or forced to serve in armed forces, and affirmed other UNHCR programs, including an emphasis on aiding refugees through development.

Another December 15 consensus resolution called for support for South African refugees. Other texts praised the International Conference on Indo-Chinese Refugees and the May 1989 Guatemala City conference on Central American refugees. Still other texts adopted that day by consensus called for aid to refugees and displaced persons in Malawi, Djibouti, the Sudan, Somalia, Chad, Ethiopia, and student refugees in southern Africa. These resolutions help draw attention to pressing refugee needs, but they do not always translate into sufficient resources or political will to solve refugee crises.

In the face of this ambivalent world attitude toward refugees, the United Nations has tried several strategies. Many refugees seek aid in already impoverished countries barely able to cope with their own peoples' needs. Malawi is home to about 720,000 exiles fleeing brutal civil war in Mozambique. Ethiopia, beleaguered by its own civil war, houses 705,000 refugees from Somalia and Sudan. Sudan shelters 690,000 Ethiopians, Chadians, Zaireans, and Ugandans. In this regard, several General Assembly resolutions in 1989 stressed the need to link aid to refugees to general development strategies, recognizing that only by aiding poor countries as a whole can one aid the strangers in their midst.

On another tack, the United Nations is attempting to develop early warning systems designed to predict when a large group of people may flood a country. This effort would allow agencies to meet exiles' needs quickly while laying the ground for their proper reception. In southern Africa, UNHCR has prepared emergency preparedness profiles for Malawi, Mozambique, Botswana, Lesotho, Swaziland, and Zimbabwe.

But at base, most refugees' decisions to flee their homelands stem from political and social crises in these homelands; only the ending of these crises will convince most to return home. Sometimes the United Nations tries to nudge refugees to return in the hope that this will help move a nation toward settling its conflicts. This has been the effort with $ million Afghan refugees in Pakistan and Iran. Operation Salaam has mounted a major effort to prepare refugees for their return home and the rebuilding of their country, hoping this would goad Afghan belligerents toward a settlement of the nation's civil war. But critics say that few refugees have chosen to return, adding that the United Nations may be overstating its success in this area.

Only peace can convince exiles to go home. A settlement of the Nicaraguan civil war-if it sees the return of contra rebels under U.N. auspices, as called for under regional peace treaties-would prove this simple thesis. In the expectation of a successful conclusion to the peace process, the U.N. Secretary-General, the U.N. Development Programme, and UNHCR called an international meeting on Central American refugees and displaced persons, June 27-28-a follow-up to the May 1989 conference in Guatemala City. Their hope was to gather support for 59 projects to resettle, settle, or otherwise benefit Central America's 2 million refugees [U.N. press release C/27, 6/26/90].

Meanwhile, the situation for most refugees will remain delicate as long as resources are scarce and nations fail to address the gap between their stated humanitarian goals and actual policies.

6
Trade Routes and Centers

Trade centers

Before the time of bridges, trade routes on land sought shallow crossings, or fords. If a river was generally wide, crossings were sought at narrow points and ferries were installed. Both ford cities and ferry cities eventually get bridges, which become economically feasible to build because of demand for their use. The existence of bridges in turn tends to focus trade routes. The city with the earliest bridges not surprisingly tends to grow faster than other towns along a river.

Many cities develop where a cargo or wagon load of goods must be broken into parts, or where it must change modes of transportation-that is, where the handling of goods must be done frequently and at the same place. A permanent settlement of persons is needed to handle the break in bulk.

A city that has grown by doing this service is called a break-in-bulk city. All ports naturally are break-in-bulk cities; if goods are stored in large quantities in a port city, or for a period of time there, the city may be called an entrepot. Break-in-bulk cities also tend to grow where trade routes cross rivers. If a river is navigable, it becomes a route in itself and gives added importance to a growing city.

Head-of-navigation cities are a type of break-in-bulk city since transfer from river steamers and barges to land transport occurs there. Of particular note is the string of head-of-navigation cities which grew up along the Fall Line. In colonial America. The Fall Line is where the streams flowing over the hard crystalline rock of the Piedmont end in falls upon coming to the softer sedimentary rocks of the coastal plain. The Fall Line cities include Philadelphia, Baltimore, Washington, D.C., Richmond, Raleigh, Columbia, and Augusta. A valley-mouth town may originate where a valley route enters a plain, and river delta towns are epitomized by Alexandria, Egypt. A railhead, the terminus of a railroad, is another example of a break-in-bulk city. When produce from the surrounding countryside is brought by truck or to a railhead, a small city springs up.

It often happens that a populous center that is some miles from the sea will become a focus for water-borne trade. A port will then develop to serve as an outport for the existing city. A classic combination is Athens and its outport Piraeus. It may happen that an existing port handling shallow-draft ships will develop a need for a deeper harbour.

Another area nearby which has an adequate harbour may become the outport for the existing but inadequate port. For example, Osaka-once a major entrepot of trade-depends upon Kobe, which was developed for handling Osaka's deep-draft modern shipping.

Rarely does a single port become the center of a trade empire, though perhaps Venice was such a port from A.D. 1200 to 1600, just as London was from 1500 to 1900. Most often, trade activity is carried on in several ports in a developed area. The trade of the Phoenicians, which reached to Gibraltar and north to Britain, centered on several ports in the Levant.

Similarly, the Baltic Sea was once the center of a thriving trade combine, the Hanseatic League. After the formation of the Hanseatic League in 1256, the Baltic Sea became a major focus of trade. Seaports, chief among which was Lubeck, were the first members of the League, but river ports soon were admitted to membership. Most cities on or at the mouths of rivers draining into the Baltic were Hansa trade centers. The end of the 14th century saw the height of the League's influence, but it was another century before its power over commerce was broken.

Trade routes

In the era before large-scale transportation by railroads and ocean-going liners, roads and canals were the chief routes for trade. The roads of Britain in the early part of the Industrial Revolution were poor in construction and suffered from competition from canals.

In the United States, canals were somewhat late in coming into popular favour, and a series of turnpikes was established in the years between 1794 and 1910. In 1794 the Philadelphia and Lancaster Turnpike was begun by a company authorized by the Pennsylvania Legislature to build the road. Other turnpikes were subse-quently built, and Ion private turnpikes were found in most of the seaboard areas.

However, since their service was inadequate, Congress in 1806 authorized the National Pike, which ran from Cumberland, Maryland, to Wheeling, West Virginia, by 1818. Westwar emigration was vastly aided by the National Pike, and for this reason it was gradually extended, reaching Vandalia, Illinois, in 1852.

Most of the turnpikes were covered with about 15 inches of pounded stones or gravel. By and large, the turnpikes in the United States were inadequate to the demands of traffic and were hard to maintain. Toward the last decades of the Industrial Revolution in Britain, the road builder McAdam began to build roads which did not require the heavy stone, hand laid foundation inherited in principle from the Romans.

According to McAdam, if a road was protected from water under its surface and from infiltration of water from above, a series of broken and compacted layers would form an adequate roadway.

While McAdam was getting his ideas accepted in Britain, the turnpikes in the United States began to suffer from competition from canals.- The canal boom in the United States lasted from 1810 to 1840. After 1840 the canal traffic came to be dominated by a growing network of railroads.

The railway boom lasted from about 1840 to 1890, during which time road building languished and vehicles using roads made little speed and often broke down from the roughness of the terrain. It was only in the 1890's that some of the eastern states passed laws for state aid to road building. This was in response to the growing demand for highway use by steam-driven cars and later by internal-combustion automobiles. Road materials gradually improved.

William Hobson took out a patent in England in 1827 for a concrete road which was to be built by methods invented by the Romans, but the extension of the principle of concrete roads did not come about until after the invention of Portland cement in the 1850's. A second method of building roads required the use of asphalt. The use of asphalt in England languished until 1869, when a part of a London street was paved with the material; the success of this paving insured a rapid spread of asphalt in England.

Rivers and river-fed canals

Any navigable river may become a trade route if it connects areas of unlike resources. The most famous example is the water route followed by the Scandinavian tribes when they found that the Dnepr River provided access to the Black Sea and to trade with Byzantium. From the middle of the 9th century to the middle of the 13th century the rivers between the Baltic and the Black Sea were used to take to Byzantium the slaves, furs, and honey of the north and to return with manufactured objects, art works, and weapons. Canals are most often dug to join navigable water bodies that are close to each other. Seldom are canals for navigation dug over long distances of land, but frequently they parallel rivers, or use river water.

The construction of canal locks to overcome differences in water level had well advanced by 1400. The first canal to have locks overcoming differences between level of water and level of -land was built in Germany in the closing years of the 14th century, between Lauenburg on the River Elbe and Lubeck on the River Trave. The next lock canal of importance was in northwestern Italy, in the area of Lombardy. This canal had originally been designed for irrigation, but it was enlarged about 1270 and used by boats. The canal ended in a basin near the western wall of Milan.

In the 1380's plans were afoot for building a new cathedral at Milan, and improvements in the canal were made. The first pound lock in Italy was constructed in this canal in 1438, and the second lock was added a few years later. These works set the stage for a new era of canal building. Canal building in Italy benefited, as did so many other fields in mechanical arts and practical technology, from the work of Leonardo da Vinci. About 1492 he began to make plans for the construction of six new locks on the Milan Canal. These were completed in 1497. After 1503 Leonardo made plans for a canal to lead from Florence to the lower River Arno, but this work had to be abandoned because of its technical complexity. By 1595 complete proposals for aqueducts, bridges, locks, and so forth for the canal between Pavia and Milan were completed. The work was started, but it was abandoned when Meda, the engineer, died.

The first important lock canal outside of Italy was the Brussels Canal, constructed in the decade 1550 to 1560. This ran from Brussels about 20 miles cast to the town -of Boom on the Rupel River and cut in half the distance that boats had previously travelled. Flanders became a center of canal building, and a 44-mile canal dating from about 1622 linked Bruges with Dunkirk. An extension was made to Ostend in 1666, and a tide lock was constructed there three years later. In England the era of canal building corresponded almost exactly to the period of the Industrial Revolution, 1760 to 1830. Its geographic effects were quite marked. It reduced the fear of local shortages of fuel and food.

It brought new agriculture into existence by bringing formerly unused areas closer to market, and it quickened economic life in the older but more removed farm areas. It also lessened the cost of transporting such bulky materials as building timber and building stone, CWT for Potteries, and iron ore and coal for new steel mills.

Summit-level canals

In France two, summit-level canals were built in the 17th century. The Briare Canal led from the River Loire northward in a gentle arc to the town of Montargis on the River Loing, which flows north to the River Seine. Construction began in 1604 and for the next 30 years proceeded intermittently; in 1642 the canal was ready, for traffic.

After its opening in the middle of the 17th century, canal operations became very profitable, and the annual trade was about 200,000 tons. Still in operation, I the canal carries annual traffic that is not far X short of one-and-a-half million tons. The second summit-level French canal was the Orleans Canal. Built between 1682 and 1692-1 to supplement the Briare Canal on the south, this canal had a length of 46 miles. Farther to the south, work was begun on the Canal du Midi in the winter of 1666 to 1667, and two years later more than 8,000 men were working on it.

Aqueducts carried the canal over the River Cesse. The length of the canal from Toulouse end (east of Carcassonne) was 150 mile 100 locks intervening. Modelled aft older Briare Canal, the Canal du Midi with) its many improvements became the Prow type for later great canal-building projects in the rest of Europe.

The third summit-level canal in continental Europe, 15 miles in length, was opened in 1669 in Germany. This was the Friedrich Wilhelm Canal from the Spree River on the west to the Oder River on the east. The canal system in northwest Germany was extended between 1743 and 1746, when a link was made between the Elbe River and the Havel River via Lake Plaue. In Sweden the canal link between Stockholm and Goteborg had been completed in 1610.

The first summit-level canal in Britain as built in the period 1737 to 1745 and from Newry to Lake Neagh in northern land. Its purpose was to bring coal from enemies near Newry. Until the beginning the Industrial Revolution (about 1760), most of the internal navigation in England via canalized rivers, supplemented over short stretches by dug canals. This system, added to by the new canals built after 1760, has been called the arteries of the Industrial Revolution.

River-joining canals

Canals joining rivers are quite common in northern Europe, where the northwardrending rivers are joined by an east-west anal system. There are many canals which run parallel to rivers and utilize their water but have separate channels. The Erie Canal, opened in 1825, used for the most part the water of the Mohawk River. Though as the New York Barge Canal it presently carries little freight, it has become important for water supply. The St. Lawrence Seaway canals parallel the river of the same name, and canals in Europe parallel the rivers Rhone and Rhine. In the United States, lagoons behind sand bars are joined to form the Intra-coastal Waterway. The Gulf portion is very important in the transport of oil by tanker, and the Atlantic portion is used mainly by pleasure craft. The early canals of Ohio and the modern canal system of the U.S.S.R. are examples of canals across portages. Moscow has been called by Russians "the port of the five seas" since the completion of the Don-Volga Canal in 1952.

The Grand Canal in China, from Peiping to Shanghai, is a rare example of a long overland canal devoted mainly to transport. It was finished in the early years of the 7th century, but it represented, in part, canals that dated back 15 centuries. It was used intermittently until about 50 years ago.

The Communist government of China in 1954 announced plans to refurbish the route. The utility of the Grand Canal is shown by its other popular name, the Grain Transport River. Stretching about 1,200 miles from north to south, it connected the main economic areas of historic China. About 150 miles of the Grand Canal are not operative near the Yellow River and are being rebuilt. The southern end connects with the Shanghai delta area, which has nearly 1 0,000 miles of locally used canals.

In Russia, Peter the Great abandoned his attempt to have the Volga and the Don joined by a canal at the point where they come within 48 miles of each other. However, he had a canal dug joining the Neva River at St. Petersburg with the Volga by way of Lake Ladoga. The draft of this canal was only 18 inches, but by 1760 more than 3,000 boats of about 80 tons each were using it annually.

A much more direct link between the Neva and the Volga was made in 1808 when the Tikhvin Canal was constructed. Another canal, the Mariinsk, was opened in 1811 and joined the two rivers via Lake Onega. This canal was designed for boats of more than 160 tons. In 1805 the Black Sea was joined to the Gulf of Riga via a canal between the Dvina and the Berezina rivers. By the middle of the 19th century, Russia was hauling barges over more than 50,000 miles of navigable waterways, The U.S.S.R. at present is well provided with canals linking most of the rivers of European Russia.

The canalization of rivers proceeded rapidly in 19th-century Europe. The Danube was worked upon from 1849 to the end of the century, at which time it had almost 2,000 miles of navigable waterways. For many decades after 1850, navigation of both the Rhone and the Rhine rivers was difficult because of the winding channels and the speed of the current in some sections. In 1840 the Canal de Miribel was exonstructed south of Lyon.

Technical problems of sitting downstream and of scouring upstream long hindered adequate barge navigation. By far the greatest amount of canal building just before, during, and after the Industrial Revolution up to the rise of the railroads was in the area of France, Belgium, and the Netherlands. In 1793 the Canal du Centre was built connecting the Loire River with the

Saone. This canal completed the first inland route from the Mediterranean Sea to the English channel. The canal connecting the Rhine the Rhone rivers was completed in 1834 and provided the second link between the Mediterranean and the North Sea. In the decade between 1818 and 1828 three major canals were built connecting various coal fields, and in 1828 the Sambre River was canalized. For many decades the Dutch strongly opposed the building of canals and the canalization of rivers in Belgium be cause of their fear that they would lose the trade showing into the Du:ch ports on the North Sea.

Grounds for these fears, of course, existed, but the general geographical in position of -the entire North Sea area was so favourable that all ports grew and canalization of rivers in one area eventually redounded to the benefit of other ports located nearby. The Dutch in 1825 constructed a 50-mile-long canal bypassing the Zuider Zee seacoast. The canal was 18 feet deep and 40 yards across, and it represented a masterpiece of canal building across very soft land. The opening of the Sault Ste. Marie Canal in 1855, with enlargements in 1895, 1896, and 1919, permitted the iron ore of the Mesabi Range to be transported cheaply and in great quantities to the ports of the southern Great Lakes.

Chicago and Gary came into being, and steel mills occurred in an almost continuous band from Detroit through Toledo to the eastern end of Lake Erie, where coal was available cheaply from the Allegheny and Kanawha plateaus to the south and southeast. This was an extension of the American manufacturing belt. The marketing of steel from the manufacturing belt was hindered because the steel had to pass by rail or via the New York State Barge Canal or the old Erie Canal to New York City. It was obvious that access to the sea could be had more cheaply if the St. Lawrence River between Lake Ontario and Montreal could be canalized by the bypassing of rapids.

The history of the development of the St. Lawrence Seaway, finally opened to deeper-draft ocean traffic in 1959, started in 1821. At that time the Lachine Rapids were bypassed by a five-foot-deep canal, which was deepened within two decades to nine feet.

Four other major rapids were bypassed by canals at about the same time, and between 1834 and 1847 the Cornwall Canal and the three shorter Williamsburg canals were completed. The route could then be used by ships of moderate draft as far as Lake Ontario.

The old Welland Canal was constructed between 1824 and 1829 to bypass the falls at Niagara and thus join lakes Ontario and Erie. The amount of trade passing through the St. Lawrence route before the opening of the deeper seaway was considerable and has grown rapidly since 1955. Irrigation canals generally do not carry freight since most of them are in dry areas having adequate -road or rail systems.

Ocean routes

In this modern age many people consider the sea to be somewhat outmoded as a highway. Nevertheless, the traffic in heavy goods between the countries of the world still utilizes the sea to a great extent.

Most of the oil, coal, ore, lumber, and grain that is transported moves across the various oceans in modern, efficient freighters, barges and tankers and thus takes advantage of the lowest possible costs. This use of the sea continues despite the expansion of air transportation to accommodate people, mail, machinery, and many kinds of articles with selling prices high enough to defray the high cost of air transportation.

Great sea routes

The great sea routes are not essentially different today from what they were in the days of sail. These routes were developed along the lines that most readily connected the various places it, the world which had different commodities to trade.

Since some areas have now a surplus of raw, heavy, cheap commodities, exchanges are made between these areas and those that have a surplus on: relatively high-cost manufactured goods. Because of the tendency to ship goods at the lowest possible cost where speed is not a factor, the ocean routes continue to handle greater and greater quantities of materials.

There is the advantage that ships carry tonnage far exceeding that of any train. Finally, the route is level: no mountains or plateaus must be crossed. In spite of the relative cheapness of sea transport, man still seeks to make the route as short as possible.

Ocean trade routes follow the shortest possible paths between the major production areas. The ocean routes have been somewhat modified in modern' times by the construction of short sea-to-sea connections, especially at Suez and Panama, and by the recent navigational improvements in inland waters, such as the St. Lawrence Seaway.

Modifications have been made in order to provide water routes for cargo as far inland as possible and thus reduce the cost of transportation. Unlike freight, whose means of transportation is governed largely by costs, passengers and mail tend to use air or land transportation as much as possible.

A freight cargo bound for London, for example, would go via the Thames Estuary, but passengers and mail would leave the ship at Southampton and complete the journey by rail, thus saving several hours. Some of the other improvements affecting ocean transportation involve the construction of artificial harbours, such as the excellent and spacious harbour of Los Angeles.

The various modifications made in ocean routes during the last century have not profoundly affected the general character of ocean transportation. Grain still moves from eastern Canada and the United States to northwestern Europe by freighter, while it moves in increasing quantities from the western coast of North America to Japan, India, and Pakistan.

Iron ore still moves from northern Chile to the steel mills on tidewater in eastern North America. Coal, which normally is not shipped long distances now that oil is widely used to power ships, is still barged in great quantities over water from the North Sea, through the English Channel, and along the western coast of Europe into the Mediterranean, as it is shipped from the "have" countries, particularly Germany, to the "have-not" countries in southern Europe.

The traffic in oil follows the discovery of new fields and the changes in consumption habits in different parts of the world. To counteract Europe's oil deficiency, tankers move in from North and South America and southeastern Asia with both petroleum and refined products.

Oil is shipped from the South American areas to the United States for refining and transshipment. Of all the methods of transporting oil, the ocean-going tanker is the cheapest, on a cost basis. The tanker not only has no rival in ocean crossings; it also competes successfully in coastal shipping. Many years ago large pipelines were constructed from the Texas oil fields to the vicinity of New York City.

Although these pipelines were used to full capacity during World War II because of the enemy, submarines which menaced the tankers, the tankers resumed their dominance at the close of the war and the pipelines were abandoned or converted to other uses.

Nodes

So far as trade between Northern Hemisphere points is concerned, four major concentrations of ocean trade routes, are important: Suez, Panama, Singapore, and the English Channel. The Suez node is of vital importance to Europe, for oil from Iran and the Persian Gulf sheikdoms passes through Suez on its way to markets in western Europe. The vital importance of this node is indicated by the history of attempts to provide a Red Sea-Mediterranean link.

The first canal connecting the Nile and the Red Sea was- constructed about 2000 BC. and remained in use for a thousand years. After that other canals were dug and used, but political changes did not permit continuous use and each canal eventually silted up. The present link was opened to traffic in 1869 and has repeatedly keen widened and deepened for larger vessels. Even now, dredges must be always on hand to remove the windblown sand which tends to fill the channel.

The temporary closing of the Suez Canal owing to the June 1967 hostilities forced oil tankers to use the route around the southern tip of Africa, and the canal continued to silt up while dredges were not permitted to operate. The opening of the Suez route had a tonic effect on the growth of Asian ports by bringing them, in practice, twice as close to Europe as before. The effect on the Asian countries was to change their focus of trade from an internal or centripetal bias to an external or centrifugal orientation.

Almost all of the major Port cities of Asia had their founding or period of important growth after 1870. These cities include Karachi in Pakistan; Bombay, Madras and Calcutta in India; Rangoon in Burma; Singapore; Saigon in Vietnam; Hong Kong, V Canton, and Shanghai on the China coast, and Yokohama and Kobe in Japan. The greatest of these as a commercial center and entrepot was Singapore.

The Panama route was opened to traffic in 1914 and had the effect of reducing the overland rail-freight rates through competition and of encouraging the rapid growth of cities on the West Coast. Los Angeles, being the southern most large city there particularly benefited.

The river

Very early the Vikings used this route and portaged their ships across a watershed 10 miles wide. By the year A.D. The small-craft -canal was in use from Lilbeck to the Elbe River. The forerunner of the present link was the Eider Canal, cut in 1784 to join Kiel with their Lakes and then the River Eider leading into the North Sea. The canal was 10 feet deep. Through these four foci of traffic, and through ones of lesser importance, moves about a fifth of the world's ocean freight on unscheduled cargo ships called "tramps."

7
Languages and Dialects

A person travelling by foot from village to village across the north Indian plain would sleep each night in a village where the speech of the neighbouring villages was well understood. Yet in the course of several days' journey, persons in the eighth or ninth village would not readily be able to understand a person from the first night's village. They would say that the other person spoke a *dialect.* If no words in common could be found they might agree, through an interpreter, of course, that they spoke different *languages.*

Language families

There is little firm agreement among linguists as to what constitutes a dialect, what is a language, and what tongues should be included in a language family or stock. A language family consists of now mutually unintelligible languages, which at one time were more closely related, and for which a common ancestor language can sometimes be inferred and partially reconstructed. Similarity in structure is one clue to relationships, and a certain amount of traceable common vocabulary another clue.

Languages

No one knows exactly how many languages are spoken over the earth, and any estimate of the number of dialects would be purely a guess. General agreement exists that there are at least 2,000 separate languages spoken today. However, only 134 are spoken by at least a million persons. The remaining languages are obviously very important in local areas, but are unknown and generally ignored by the outside world. For instance, a census in 1961 showed that in Scotland only 1,079 persons had Gaelic as their only language.

Languages as structural systems

No language spoken by any group is gibberish, though it may sound so to the outsider. Almost every child in its babbling stage produces every sound used in every language. As he listens to his elders, he be- gins to pay attention to those sounds which he hears them produce. Rewards follow correct reproduction of their sounds, and disapproval follows if phonetically near but incorrect sounds emerge from the baby's throat. Thus a mother tongue is learned. If the child is raised in a family speaking American English, he soon disregards his ability to produce the Arabic *hamza,* the nine Cantonese tones, and Hottentot clicks. Attitudes toward sounds not used in one's mother tongue vary with age if a person remains monolingual. When he is young the sounds are merely unrewarding. As he grows older, the sounds by turns become hard to imitate, impossible to produce, and finally "evil." Just as sounds are not gibberish, so are the words of a foreign language not random and unstructured. Each language has a finely structured way of expressing thoughts, and many of these are quite euphonious.

The Semitic languages, which include Arabic, Hebrew, Maltese, Amharic, Aramaic, and ancient Akkadian, have generally a *three-consonant cluster* as kernel for vocabulary. By various use of vowels, prefixes, and suffixes the basic kernel may express a large variety of related ideas. In each case, the order of the kernel elements remains unchanged. The kernel *k-t-b* in Arabic, for example, refers to a book or something written. Table 1 lists a dozen words using the k-t-b kernel.

Table 1
The k-t-b Kernel

Arabic	*English*
kitaab	book
maktuub	letter
yiktib	write
kaatib	writing
kitaaba	writing
kutby	booksellers
kutabhana	library
katatyb	day-school
katab	note
makaatib	writing table
iktibaar	research
muktabar	laboratory

The word or *day-school* above illustrates another characteristic of Semitic languages: intensification. When an idea is to be stressed, the middle consonant of the kernel is doubled: a day-school is a place where much writing is done. The mental capacities of men everywhere, and for the past several millennia at least, are much the same. Each group contains a few dullards, a scattering of geniuses, and the great middle range of average intellects. Owing to this condition, we cannot say that there are any primitive languages. Indeed, languages spoken by groups with a low level of material culture are amazingly complex and expressive, as indicated in Eugene Nida's superb and delightful *Linguistic Interludes.*

Each language is adequate to the situations of its speakers. If their life should become more complicated, the vocabulary of the language would increase accordingly. Important things to the culture are represented by many special words, and unimportant things slighted, It is said that Arabs have 80 words for camels in various manifestations, and that Polynesian tongues have many words to describe in fine graduations the blue-green colours of sea, sky, and vegetation.

Examples are numerous. Bedouin Arab culture pays as much attention to camels as modern American culture does to sports. Travellers often report on such vocabularies. Leslie C. Stevens writes in *Russian Assignment:*

> ... there were long stretches of road where the continual wind had swept away the snow and left nothing but bare, slippery ice. The Russians have a special word for that sort of ice. As might be expected, the language is very rich in words for all sorts of ice and snow and frost, as it is for varieties of horses, different aspects of drunkenness, and for the family relationships that we have to express with hyphens in English.

Bilingual persons and high-low speech

Areas where a considerable portion of the population regularly can command at least two languages as occasion dictates may be considered as areas of bilingualism. Reasons why an area has bilingual speakers include recency of migration (as Indians in Fiji, East Asians in Hawaii), linguistic pride (as Germans in south Chile), absorption of former isolated groups into a national economy (as Kurds in Iraq, Amerinds in Central America, Quechua and Aymara speakers in Peru, and Eskimos in Arctic North America), colonial residues (use of English in Africa and Asia, use of French in West Africa), and encouragement of minority-group languages (as practiced in the U.S.S.R.). In some instances, a combination of these reasons applies to bilingualism of an area. Ability to shift one's linguistic habits as the occasion arises is widespread over the earth. Perhaps the United States is the largest example of a country where most of the people command easily only one language or major speech form. There are two ways of looking at "two-language" ability. Where a person commands two separate languages, we say that he is *bilingual.* Where the cultural and social situation demands that one sphere of life be discussed in one way of speech, while another sphere be the domain of a different way of talking, we say the situation is one of *diglossia.* Different sets of values, attitudes, and behaviours are called for when using the two modes of speech.

These modes are generally divisible into High and Low, where the social feeling is one of respect in the first, and one of intimacy in the second. Diglossia and bilingualism are widespread. Where the High and the Low speech forms are from the same language stock, the situation is one of simple diglossia without bilingualism. Contrasting uses of High and Low languages.

Simple diglossia occurs rather widely. In the German-speaking areas of Switzerland, the language of education is standard German, whereas the language of the home and street is Swiss German. Two language variants are found in Norway. In Arab countries the language of the mosque, of education, and of poetry and newspapers is classical Arabic, while the language of the street, the schoolyard and the coffee house is colloquial Arabic. The rulers of Arab countries can communicate only in classical Arabic, as the colloquial Arabic spoken in each country is distinctive and cannot be understood readily by visitors. Each nation feels that its own colloquial speech is closest to the classical forms used in the Koran.

In Okinawa the language of the school and of the

newspapers is standard Tokyo Japanese, while the Okinawan language, though related to Japanese in the distant past, is used at home and in the markets. In Greece the High language of newspapers, government, and literature is *katharevusa*, derived from classical models about 1820 A.D. In the markets, and for much of modern poetry, people use the spoken *dhimotiki*. Each language is thought to have its proper sphere, and when the New Testament was published in the popular language in 1903, there were riots throughout Greece.

True Bilingualism

Several areas of the world have in the past had a truly *bilingual* situation, with little real communication between groups speaking the two languages. In pre-1917 Russia, many of the elite preferred to speak only French, while the masses spoke Russian or other Slavic tongues. In southern India even today, many elite people prefer to speak in English, while the common people speak local languages. The reason the elite prefer English is that they feel that Hindi, the designated national language, will undercut the use of their local languages. In much of western Africa, the elite speak the language of their formal education, which is English for former British territories, and French for those formerly allied to France.

Table 2.

Uses of high and low languages

Uses of Low Language	*Uses of High Language*
Instructions to servants, waiters, clerks, workmen	Sermons in church, temple, or mosque
Conversation with family, friends, colleagues	Speech in parliament, personal letter, university lecture
Radio-television "soap opera"	News broadcast
Caption on political cartoon	Newspaper editorial
Folk literature	Poetry

Personal Bilingualism and social diglossia

It sometimes happens that two totally separate languages are used for the High and Low situations. This is a combination of personal bilingualism with social diglossia. Such is the pattern in Paraguay, where Spanish is the language for speech related to education, government, and religion. An Indian language, Guarani, is used for intimate and home-based talk.

Linguists report that only about 4 per cent of the people are monolingual in Spanish. About 56 per cent of the people are bilingual, while about 40 per cent—mainly residing in the more remote portions of the countryside—are monolingual in Guarani. Gregores and Suarez report on the use of the two languages:

> The upper classes tend to use Spanish more frequently than the lower classes, and people in towns more often than the rural population. Likewise, any person in a socially "inferior" position with respect to some other person would not, as a rule, address him in Guarani, unless the latter has already done so; in many cases, this pattern may extend to the parent-child (older relative-younger relative) relationship Among "equals," Guarani is the language of intimacy, of love, and friendship; Spanish the language of respect. A stranger would be ad dressed in Spanish if he looks respectable; other wise Guarani would be used.

One other of many possible examples of High-Low use of unrelated languages is the use in pre- 1914 eastern European Jewish communities of Hebrew and Yiddish, the latter a mixed language largely German in Syntax and vocabulary.

Trade languages, Pidgins, and Creoles

As people move willingly or unwillingly over the face of the earth and come into contact as large groups with persons speaking quite different languages, the stage is set for the development of hybrid speech forms. The fraternization jargon developed by soldiers is both amusing and familiar to many Americans. Most such hybrid languages die out with decline in contact between the groups. However, one wartime hybrid survived as a language of trade: *Lingua Franca* was a military hybrid tongue of the Crusaders and Saracens. The term has become descriptive of trade languages in general. The conditions most favourable for development of contact hybrids occur when slaves or contract labourers with different mother tongues are forced to try to communicate with each other in the language of work supervision. In the early stages of such hybrid communication, the vocabulary shifts rapidly and there is much flexibility in use from decade to decade.

Simplification of structure and vocabulary of the main language used in the work supervision situation occurs as the hybrid comes to be used for nonwork activities. Where there is no strong social stigma against using hybrid speech, and where no correction of hybrid usage occurs, a *pidgin* language arises. Pidgins often incorporate vocabulary and basic grammar from the prestige language within a simplified sound system and word order framework of the worker's first-learned language. When the first plantation workers came to Hawaii from East Asia, they spoke a variety of languages and dialects. The *lingua franca* was English, and in the course of being simplified for use on the plantations, became a mode of speaking called pidgin English. In most rural and in some urban areas of Hawaii, pidgin is the main language of children, used in the street and the schoolyard. In areas such as Haiti and other Caribbean areas, pidgin tongues have become mother tongues of large groups, and the vocabulary through the decades greatly expanded. When a pidgin creatively expands in this manner, it is called a *creolized* language. Hawaiian pidgin started to become a creole in the 1920's and 1930's, but with the advent of television it is gradually disappearing, and will not pass into the full creole stage.

Diffusion of Pidgins

The word "pidgin" is a 17th-century English variant of a South American Amerind term *Pidian,* meaning "trader." Many cultural geographers and anthropologists prefer to seek a diffusion explanation for two similar phenomena in different places, rather than attribute the occurrences to independent invention. The anthropologist Dell Hymes with regard to pidgins notes:

> One major theory is that almost all the world's pidgins and creoles have their origin in an Afro-Portuguese pidgin developed on the coasts of West Africa, itself perhaps adapted from a Portuguese version of the medieval Mediterranean pidgin, Sabir, and subsequently rapidly replenished in vocabulary *(relexified)* from Spanish, English, Dutch, or French, as the case might be, in various parts of the Caribbean, the Pacific, and Asia.

Distribution of the main trade languages, creoles, and pidgin languages is shown in Figure 23.3, as compiled by the linguist Robert A. Hall, Jr.

Dialects

Almost everyone has had the experience of speaking with a countryman who consistently pronounces words differently, perhaps with a different lilt and speed. Others may use different words to describe a common object. Within a language there are often such regional ways of pronunciation and such regionally distinctive words. Only rarely do phonetic and vocabulary boundaries coincide, but both are studied by dialect geographers.

Pitch accent

Sound shift or phonetic change studies delineate the changes from place to place in vowels, consonants, and the lilt of a spoken word. Results of very detailed studies on contrasts in pitch accent of Japanese words.

The making of vocabulary boundaries

It takes years of careful field study by numerous experts to gain scientifically adequate dialect information. After the careful gathering of dialect data from persons born near the place of interview comes the delicate art of drawing dialect boundaries. When boundaries for a great number of separate word usages are drawn, the boundaries are called *isoglosses.* Where many isoglosses bundle together, linguists feel confident in drawing a dialect boundary. The work of dialect geographers accomplished by one man, nor in a single season. Many field interviews and accurate records contribute to the final map.

The boundaries finally assigned to the dialect areas of the eastern United States derived from thoughtful considerations of both dialect data and historical migrations. In a few years it may become possible to draw equivalent dialect boundaries for the Mississippi basin and the Great Plains. From the Ohio map, the Texas map, and other pieces of evidence it is clear that the northern forms disappear into Canada near Great Falls, Montana, while the southern forms terminate at the Rio Grande near Laredo, Texas.

Language change

The speech of each generation of a family is so close to that of other living generations that the language appears the same to all speakers. However, children's slang words may be new to their grandparents, who in turn occasionally may use a word not a part of the children's vocabulary. The only *broom,* for instance, known to

American children today is a sweeper, whereas its homophone *brougham* was a familiar word in the early 1900's for one style of automobile.

Over a period of several centuries such small and rarely noticed shifts in word use accumulate, and the vocabulary of the language changes measurably. Linguists do not agree on whether all languages change at the same rate. However, a basic vocabulary decay rate of about 81 per cent per thousand years has been found for a number of related-language sets. It is sometimes possible in a country of a single language to trace the changes in an item of vocabulary.

The newest word is the most common, but centers in the area of the old cultural hearth around Kyoto and Nara. As the older words are encountered still in use, one gets not only farther from the cultural hearth but finds ever fewer examples of relict words. Dialect areas tend to be the more isolated portions of a country, where communications in the past have been hard, and barriers many. Mountain ridges often, though not always, mark dialect boundaries, making valleys and small drainage areas little enclaves of older speech forms. Islands which have little communication with nearby mainland areas often preserve archaic forms and may also have a slightly different vocabulary and variant stress in speech.

The Korean spoken on Cheju Island, for instance, is totally incomprehensible to mainland Koreans. Within a language the neutral and frequently occurring words are much less likely to have regional variants than are family terms of affection used in the aggregate much less often. The four words for interrogative "who" in Scotland stand in contrast with the fifteen words for "youngest or smallest of a brood or litter."

Very ancient words for agriculture and for natural objects are often similar through many adjacent dialects and languages. Much work remains to be done in both linguistic geography and in the more restricted field of dialect geography. Only when many studies are completed for dozens of small areas will we have an adequate look at the way in which man, the only animal with language, has imprinted the noosphere with his words and symbols.

8

Transnational Culture

Anthropologists have historically resisted engagement with human rights movements because they feel that the concept of human rights is an artefact of Western cultural traditions raised to the status of global normativity. The parallels with imperialism have been too close for anthropological comfort.

The new global discourse of human rights re-poses this question for anthropology. The simple stance of tolerance has become increasingly and in the face of feminist and human rights activists' demands for global attention to practices such as violence against women and political torture. Yet, anthropologists rightly recognise that human rights language is historically rooted in the Western legal tradition and grew out of its particular social conditions. The global spread of human, rights discourse is similar in many ways to the imperialist introduction of legal orders from the West to the rest. Does this mean that anthropologists should continue to shun human rights discourse as a new phase of Western imperialism?

In the current era, such a rejection of human rights discourse as too 'Western' is a mistake. The present global situation is very different from that of the earlier era of imperialism. There are no longer cultures for whom the legal regimes of the West are totally alien or irrelevant. Even during the high colonialism of the early twentieth century when anthropologists writing the classic descriptions of indigenous legal systems in colonised areas treated them as they were pristine, most groups had a) ready been influenced by Western law. These accounts of isolated legal' systems were artefacts of a particular era of anthropological theorising during which colonial influences were analytically subtracted. Even the partial isolation of the 1930s and 1940s is no longer found in the 1990s—the era of globalisation of AID, transnational capitalism and the Internet. Members of small, isolated communities increasingly speak in the global arena, whether they are Penang tribesmen or Kayopó from the Amazon.

And what they speak is law; a law which is a multi-layered amalgam of United Nations resolutions, national law, and local categories and customs. Human rights is obviously based on Western liberal-legalist ideas, but in the postcolonial world it is no longer exclusively owned by the West. As various societies mobilise Western law in their demands for human rights, they reinterpret and transform Western law in accordance with their own local legal conceptions and with the resources provided by the global human rights system. They talk rights, reparations, and claims—the language of law—but construct a new law out of the fragments of the old.

There are two sides to this process. One is the incorporation of local understandings, the other is the addition of global legal discourses. To some extent, the law mobilised in indigenous rights movements is becoming vernacularised, analogous to the way languages become vernacularised over time. In postcolonial states, imperial languages are typically separated into various kinds of pidgins with distinctive accents, vocabularies, styles of speaking, and grammars. The English spoken in Nigeria is very different from that spoken in India, the continental United States, and

Hawai'i, for example. In each region it has been shaped by local languages and speaking patterns. On the other hand, what is being vernacularised is an imperial system of law based on rights and procedure, just as the vernaculars listed above come from English.

Legal vernacularisation is linked to the emergence of new national identities much as the development of vernacular languages served as the basis for European nationalism. According to Anderson's analysis of the rise of European nationalism, just before the sixteenth century, administrative vernaculars separated themselves out of the wider network of Latin-using Christendom. Gradually, and by serendipity, some local languages became the language of an entire state while others remained as regional languages. Print capitalism and the spread of Protestantism in the sixteenth century further entrenched these languages as those of government and of the nation. Vernaculars entrenched through book publishing and the emergence of reading publics laid the basis for a developing national consciousness in place of the broader network of communication between educated elites based on Latin. The emergence of vernaculars eroded the transnational power of Latin as a mode of communication linking all of Christendom. As indigenous peoples mobilise human rights as the basis for conversations about justice, they typically vernacularise and globalise the law they use. The Hawaiian Sovereignty Movement, and in particular Ka Ho'okolokolonui Kanaka Maoli held in Hawai'i in 1993, exemplify the creative reconstitution of law by an indigenous group. Other indigenous groups have followed similar strategies. These movements show that the emerging regime of global human rights is not simply the imposition of Western cultural forms and legalities. Instead, human rights is an open text, capable of appropriation and redefinition by groups who are players in the global legal arena.

The concept of culture has historically been a stumbling block to appreciating the extent to which transnational cultural systems such as human rights are vernacularised. As long as cultures are seen as integrated, cohesive, bounded, and more or less static, it is simple to perceive human rights as an intrusive, alien discourse. However, recognising that cultures are complex repertoires of systems of meanings extracted from myriad sources and reinterpreted through local understandings and interests provides a more fluid way of considering how human rights might be incorporated into local cultural practices and understandings. Cultures consist of a wide assortment of signs and meanings in discrete locales. Contemporary human rights movements clearly draw on Western concepts of law and legality.[2] Since 1945, the United Nations has served as a central source of human rights, through its covenants on economic, social, and cultural rights and on civil and political rights. The principles enunciated in the UN World Conference of Human Rights in June 1993 asserted the universality and indivisibility of human rights, the right to development and to self-determination, the role of civil society in contributing to the development of human rights, the importance of more effective implementation of human rights at the international and the national level, and the need for human rights education in a systematic way and on a massive scale. Human rights discourse has gained greater importance than ever in the 1990s and the UN has emerged as a new factor in international relations, challenging state sovereignty by establishing conditions under which international intervention is legitimate.

Indigenous rights movements build on the human rights framework, which has cumulatively incorporated formulations of indigenous rights since the post-war period. Until the late 1950s the notion of indigenous rights was dominated by the International Labour Organisation and the image of the indigenous primitive who needs to be modernised. From the 1970s to the present, the discourse has increasingly been shaped by the UN and associated with an image of the noble primitive dose to nature. Thus, the meaning of indigenous rights has changed over time. Terminology has also shifted from indigenous populations to peoples and indigenous peoples have gained greater voice in the process of articulating indigenous rights.

During the 1980s and 1990s, the UN has produced a stream of documents on indigenous peoples' self-determination and autonomy, primarily through the Working Group on Indigenous Populations, established in 1982. Over the last decade it has produced a Draft Declaration on the Rights of Indigenous Peoples which proclaims the right of indigenous peoples to self-determination and to the full enjoyment of all human rights recognised in the UN Charter and international human rights law. Indigenous groups have consistently sought to shape the language of the Draft Declaration and the policies of this Working Group through participation in its meetings, presentation of reports and a wide range of other kinds of political activity. Yet, despite the extent to which human rights discourse has proved fruitful for indigenous rights movements, it has also constrained their claims and the kinds of remedies they can seek.

The Hawaiian Sovereignty Movement, a political movement by some of the descendants of the Polynesian Hawaiian people inhabiting the Hawaiian islands before Captain Cook arrived in 1778, draws on national and global law and, to a lesser extent, indigenous law to frame its demands and claims. There are many different factions among those who currently self-identify as

Hawaiians and seek to improve the economic, social, and political position of Native Hawaiians. These groups espouse an equal number of understandings of sovereignty and self-determination. The most radical factions are demanding political sovereignty for Native Hawaiians and attempting to construct a land base for this claim. International human rights law has become increasingly central to the claims of the more radical groups, transforming this movement since the 1970s and providing the framework for asserting claims and specifying redress. With this transition to the use of global indigenous rights has come a move from a 'local' constituency to a Native Hawaiian one emphasising blood and descent rather than longterm residence and the linguistic and social markers of 'local' culture.

The political movement demanding sovereignty grew out of a much larger and older renaissance of Hawaiian culture, language, arts, dance and music. The cultural movement is inclusive, drawing in a large proportion of the 'local', or long-term residents of the islands regardless of their 'blood' or ethnic heritage. The sovereignty movement, in contrast, tends to be exclusive, emphasising the claims of people deemed to be of Hawaiian descent. Given the nature of intermarriage practices over the last 200 years, however, such distinctions are hardly unambiguous. For many, Hawaiian identity is a matter of self-identification and social situation validated by some ancestral heritage. Many who did not claim Native Hawaiian identity in the past are now re-identifying as Hawaiians. Thus, the sovereignty movement is a political movement founded on a very complex politics of identity.

The claims for Hawaiian sovereignty grow out of the 1893 US invasion and 1898 annexation of the independent Hawaiian nation. During the late nineteenth century, increasingly powerful American and British sugar planters who controlled the Hawaiian Kingdom's economy sought duty-free access to the US market for their sugar. Although American and British legal and political systems had been introduced in the mid-nineteenth century, Hawai'i remained an independent constitutional monarchy until 1893. The US minister 'in Hawai'i at the time was eager to annex Hawai'i to the United States.

But the incoming US president, Grover Cleveland, balked at this seizure of an independent nation over the protest of its legitimate ruler. He sent a representative to Hawai'i to investigate the takeover. In a speech to Congress in 1893, Cleveland clearly condemned the action as carried out by force and without the consent of the Hawaiian government. However, the election of William McKinley in 1896 and the imperialist fervour of the Spanish American War led to annexation, although the extent of opposition was sufficient to prevent passage of the annexation treaty by the required two-thirds majority of the Senate. The Senate Foreign Relations Committee circumvented this procedure by introducing a joint resolution of annexation which required only a majority vote. In 1898 the United States annexed Hawai'i with the consent of the government of the Republic of Hawai'i - the name taken by the Provisional Government - but without the consent of the Queen or the Kanaka Maoli people.

On 23 November 1993, President Clinton signed Public Law 103150 which was an apology to Native Hawaiians for the American diplomatic and military role in the 1893 overthrow and its consequences for them. This Law acknowledges that the indigenous Hawaiian people never directly relinquished their claims to their inherent sovereignty as a people or over their national lands either through their monarchy or through a plebiscite or referendum.

Hawai'i remained a Territory until 1959. In that year, a majority of the eligible inhabitants, of which the Native Hawaiians were now only a small fraction, voted for statehood. There was no alternative option available on the ballot, no choice for sovereignty offered. Some Native Hawaiians worked for statehood in the hope that a shift to more representative electoral politics would give them greater control than they had under the territorial system of appointment and limited voting franchise. During the 1960s, however, it became increasingly clear that the Native Hawaiians, now a small part of the total population of the state, were declining both in political power and in their overall social and economic conditions. By 1986 Native Hawaiians were only 20% of the population of the state. Fully 96% of the population were of mixed ancestry. They were at the bottom of all ethnic groups in Hawai'i in health, educational levels and representation in managerial and professional jobs and income, while at the top in unemployment, non-participation in the labour force and proportion living in poverty.

At the same time, Native Hawaiians experienced a rapid and profound transformation of their culture and language due to decades of pressure for American cultural assimilation, suppression of children speaking the Hawaiian language (banned in schools in the 1930s), and the widespread 'orientalising' of Hawaiian culture by tourism. Beginning in the 1960s, rapid economic change, military expansion and the growth of tourism disrupted rural communities where indigenous language and practices such as taro farming and fishing were still prevalent. These dwindling areas have become a key source of 'traditional' knowledge for the Hawaiian movement.

By the 1990s, the Hawaiian cultural renaissance was widespread, diverse and generally accepted, while the political movement, with its more radical demands for self-determination drew repeated attacks and opposition from conservative political groups. The cultural revitalisation movement endeavours to recover or recreate Hawaiian language, to revitalise Hawaiian music and dance (rescuing it from its commercialised manifestations), to rewrite Hawaiian history based on sources other than the early missionary accounts which constituted the basis for standard histories until recently, to re-establish technologies of taro farming, poi manufacture, fishing, ocean-going canoe building and navigation, and to promote artistic and craft p rod uctions such as lauhala weaving, lei-making, carving, fabric-making and design, quilting, and many others. The movement includes an assertion of Native Hawaiian spirituality and connection to the land, the revaluing of Native Hawaiian healers and rights to gather medicinal plants, and a continuing emphasis on the importance of the family, the 'ohana. Campaigns to protect sacred sites such as burial grounds and heiau (temples) draw on Native Hawaiian religious beliefs as well as legal claims.

Although the cultural florescence is not intimately connected with the political movement, it reinforces the movement in its emphasis on a renewed pride in Hawaiian culture and identity. Hayden Burgess, or Poka Laenui, one of the early leaders of the sovereignty movement in Hawaii, traces the inspiration for the Hawaiian movement generally to the spread of decolonisation in the Pacific islands beginning with Fiji in 1970, the black power and American Indian movements for local cultural rejuvenation, and the impact of the Vietnam War. Music took on new vigour, hula halau or dance schools gained wider prestige and membership, canoe clubs became more popular, and interest in Hawaiian language and healing with natural medicines expanded. Native Hawaiians became more familiar with their history and began to use their personal Native Hawaiian names in public.

The current political movement grew out of land struggles in the 1970s. The effort to stop military bombing on the island of Kaho'olawe was an explicit challenge to the military. In evictions from Sand Island, Makua Beach and Waimanalo there were challenges to the jurisdiction of the courts to try Native Hawaiian citizens. The claim that lands ceded to the US government at the time of annexation were stolen was raised in a packed courtroom in the Makua Beach eviction case. These lands had been taken from Queen Lili'uokalani by the Provisional Government in 1893 and were ceded to the US over protests and petitions by Kanaka Maoli people.

The legal system was always a central arena for political claims. By the early 1990s, protests included the refusal to pledge allegiance to the US flag in schools or to stand for the national anthem, to file tax returns, to pay income taxes, or to accept the jurisdiction of American courts when charged with criminal offences. In 1978, the State of Hawaii created an Office of Hawaiian Affairs run by nine Trustees elected by Native Hawaiians. Burgess points out that this may be a unique response to indigenous peoples' demands for self-determination because the Trustees are directly elected by the indigenous people and are, at least in theory, answerable only to Native Hawaiian constituents. But, because OHA was created by the state in order to provide an organisation with which the government can negotiate Native Hawaiian claims and because it is a state agency dependent on the state constitutional government, many activists challenge its ability to serve as a legitimate voice for Native Hawaiian interests and point out that it has often defended or promoted state policies which are to the detriment of Native Hawaiian people.

Although the political side of the Hawaiian movement emerged out of community land struggles, during the 1970s it gradually shifted from 'local' people's rights against development by state, military and corporate interests to the assertion of indigenous Hawaiian rights. By the 1980s, a powerful movement drawing on the language of human rights and pronouncements of UN documents and Declarations on Human Rights was gaining increasing visibility. The movement that began as a call for restitution became a dear demand for Native sovereignty. Emerging UN initiatives concerning indigenous peoples encouraged the movement to adopt the language of aboriginal rights. For example, in her recent book, From a Native Daughter. Colonialism and Sovereignty in Hawai'i, Haunani-Kay Trask, a prominent leader of the Sovereignty Movement, asserts that the Universal Declaration on the Rights of Indigenous Peoples should become the framework for discussions of the special status of indigenous peoples. The Kanaka Maoli Tribunal recommended that this UN declaration serve as the minimum standard for negotiations between the US and the Native Hawaiian peoples. The turn toward a global definition of the struggle has coincided with an increasing exclusivity in participation—a greater emphasis on Native Hawaiian ancestry.

A major initiative of the sovereignty movement was the creation of Ka Lahui Hawai'i in 1987, a sovereign Native Hawaiian nation-within-a-nation with a constitution, officers and citizens analogous to Native American nations. Ka Lahui claims to speak for Native Hawaiian people in negotiations with the federal government just as representatives of Indian nations do now. It seeks to establish a land base through the recovery of the so-called ceded lands, those taken from the Queen, and the Hawaiian Homelands. Non-Native

Hawaiians are eligible to join Ka Lahui as honorary members.

The government of Ka Lahui Hawai'i is similar in some ways to that of the US, with a constitution and an executive, legislative and judiciary branch, but incorporates distinctively Native Hawaiian principles, language and structures such as a council of Kupuna (elders) and a council of ali'i (chiefs) whose jurisdiction covers culture and protocol.

Rights enunciated in the constitution include protection against unreasonable searches and seizures, rights of an accused, habeas corpus, and customary rights such as free access to the mountains, caves, the sea, and sites of religious and cultural importance for personal, subsistence, religious and cultural purposes. Ka Lahui is committed to self-determination for Native Hawaiian people and has enunciated Native rights to worship, fish, gather and undertake other commercial activities. Its land claims include the private trust lands of several major trusts. Citizenship, with the right to vote and hold office, is open to all Native Hawaiians while non-Hawaiians are welcomed as honorary citizens. Ka Lahui is founded on principles of peace, disarmament and self-sufficiency. Thus, Ka Lahui represents a vernacularisation of the political and legal system of the United States which is appropriated and reinterpreted in terms of Native Hawaiian culture and politics. At the same time, it makes claims to nationhood in the international arena on the basis of a -global discourse of nation: constitutions, laws and definitions of citizenship.

Ka Lahui has made repeated submissions to the UN Working Group on Indigenous Populations and has negotiated treaties with five Alaskan and American Indian Nations. One of its major objectives is educating Native citizens in land and water resource management, exploration of trust assets, legislative processes and historic sites, and working to understand the land base and trust assets and how the Western political system has disenfranchised them.

Although the Office of Hawaiian Affairs was not initially in favour of sovereignty, it came around to advocating sovereignty by 1988 as the movement strengthened. Ka Lahui leaders worry that OHA serves as a false front for the sovereignty movement, agreeing to development projects that would not benefit the Native Hawaiian people. In 1990, an agreement was reached between OHA and the governor's office to pay back rent from the income from some of the ceded lands of $100 million in 1991 and $8.5 million every year thereafter to OHA. These payments are to settle claims for the ceded lands and thus extinguish claims to the land itself. The ceded lands are 1.4 million acres of Crown and Kingdom lands seized by the Provisional Government, transferred to the US at annexation and to the state in 1959 with the provision that the Native Hawaiian peoples were to be special beneficiaries of these lands. Clearly, there are substantial interests at stake in this political struggle.

During the summer of 1993, the centennial of the coup which removed the last Native Hawaiian monarch from the throne and replaced her with a Republic dominated by a small group of whites of US ancestry, one of the most radical pro-sovereignty groups under the leadership of Kekuni Blaisdell held a tribunal to try the United States for its takeover of the sovereign nation of Hawai'i and its acts of resource appropriation and cultural destruction to the Kanaka Maoli people. Ka Ho'okolokolonui Kanaka Maoli (the People's International Tribunal of the Native Hawaiians) was a 10-day trial which moved from island to island taking testimony from experts and ordinary people before an international panel of nine judges. One hundred and thirty-five people testified.

Ka Ho'okolokolonui Kanaka Maoli claims as its ancestry about 50 tribunals of this kind over the past 30 years in which the citizens of a nation hold their government in judgement, inspired initially by a tribunal convened in 1966 by Bertrand Russell to examine the war crimes of the US during the Vietnam War.

The Tribunal hoped to draw international attention to the situation of the Kanaka Maoli people. Its stated goals were to elicit and assemble a record of the wrongs committed by the US and its subsidiaries against the Kanaka Maoli people and to educate the Kanaka Maoli, the United States and the world about these wrongs. One of the most important audiences for the Tribunal was the UN Working Group on Indigenous Peoples. According to Glenn Morris, one of the three prosecutors in the Tribunal, an important goal of the Tribunal was producing a document for submission to the Working Group. The initial statement of charges was submitted to the US President, the governor of Hawai'i, the Secretary General of the UN, and the chairperson/ rapporteur of the UN Working Group on Indigenous Populations. A statement on behalf of the Kanaka Maoli was submitted to the fiftieth session of the Human Rights Commission in February 1994 and several members of the Tribunal organising committee went to Geneva in July 1994 to present testimony to the UN Working Group on Indigenous Populations.

The Tribunal took the form of a criminal trial, beginning with a complaint—a formal statement of nine charges against the United States by the Kanaka Maoli people. The complaint was drafted by the three prosecutor-advocates, Glenn Morris, Maivan Clech Lam and Jose Morin, and edited by the Tribunal Komike (committee). These charges included interfering in the internal affairs of a sovereign people and nation, annexation of a sovereign people and their nation and territory without their free and informed consent, illegal appropriation of the lands, waters and natural resources of the Kanaka Maoli, economic colonisation and dispossession of the Kanaka Maoli, and acts of genocide and ethnocide against the Kanaka Maoli including the replacement of Kanaka Maoli institutions with Western ones and the suppression of the Kanaka Maoli language. Each charge cites the laws which have been violated. The laws included the laws of the Kanaka Maoli nation, international treaties from the nineteenth century, the US Constitution, and various UN declarations on human rights and the rights of indigenous peoples.

Among the several remedies sought in the complaint were the right to 'the reassertion and effective exercise of [the Kanaka Maoli] right to self-determination and sovereignty', the demand for negotiations between the Kanaka Maoli and the US to develop a 'just and evolving, as need be, relationship between the two parties', beginning from the premises of the rights enunciated in the Draft Declaration on the Rights of Indigenous Peoples as developed by the United Nations Working Group on Indigenous Populations. The complaint also demands the return of all Kanaka Maoli territory. Clearly, the statement of charges and the remedies sought are framed in the language of the law, but this is a plural law in which the law of the nation is nested between indigenous law and global human rights law of the United Nations.

The US, as defendant, was sent the complaint and invited to appear in the trial, but did not. Consequently, an empty chair sat beside the judges at each hearing labelled 'US Representative'. Prosecutor advocates organised the taking of testimony. Hearings always opened with Kanaka Maoli ceremonies and prayers. Testifiers were both experts and 'taro-roots' (grassroots) people. Among the participants were representatives of other peoples colonised by the US in 1898, such as the Philippines, Guam, Puerto Rico and Samoa. The Tribunal began with the testimony of several historians who are re-examining Hawaiian history and challenging the conventional history of voluntary acquiescence to American control by emphasising the violence of early encounters, the coercive assimilation of the colonial process, and the economic exploitation and resource appropriation of American colonialism, including the occupation of lands for military purposes.

As the Tribunal moved to other islands, it heard testimony in places where Kanaka. Maoli communities were contesting land rights. On each island, local people testified about problems ranging from the encroachment of hotel and golf course development, inadequate housing provided by the state Department of Hawaiian Homelands, obstacles to asserting claims to family lands, the erection of fences and other forms of denial of access to traditional gathering and fishing areas, to the destruction of natural fishing and gathering environments, and the suppression and tourist commercialisation of Kanaka Maoli culture.

The Tribunal concluded that the US was guilty of violating na kanawai, elements of customary international law, and the US Declaration of Independence. The US engaged in interventions which violated its international treaties with the nation of Hawai'i, it supported a coup against the legitimate government of the country, and after annexation it forcibly degraded and dispossessed the Kanaka Maoli. US assertions of jurisdiction and property title in the Hawaiian Islands were held by the Tribunal judges to be legally invalid. Further, the judges concluded that the inherent right of the people to sovereignty has not been extinguished and that the Kanaka Maoli have been subjected to ongoing processes of physical and cultural genocide by the state and the US government. Ten years earlier, a Congressional Commission appointed by the Reagan administration to study the situation of the Kanaka Maoli people divided on these points. A majority report claimed that the US invasion and coup of 1893 was not unauthorised and that in 1959 Kanaka Maoli had not been incorporated into the US without the exercise of self-determination while a Kanaka Maoli minority wrote a separate dissenting report arguing that Kanaka Maoli had been subjected to an illegal takeover in 1893.

The Tribunal recommended that the US and the world recognise the sovereignty and right to self-determination of Lahui Kanaka Maoli under several covenants of international law including the International Covenant on Civil and Political Rights and the International Covenant on Economic, Social and Cultural Rights. It recommended the return of all Kanaka Maoli lands including ceded lands, and associated water rights. The Tribunal recommended that in negotiations with Lahui Kanaka Maoli, the US should observe the provisions of the UN Declaration on the Rights of Indigenous Peoples as the minimal standard of negotiation.

Thus, the Tribunal appropriated Western legal forms and symbols in an effort to harness the power and legitimacy of conceptions of law held by the US government, the UN and the people of the United States. The Tribunal took on the form of a trial, of criminal legality, without the sanctions or the authority of a state. This was a process full of irony. As one of the organisers asked, how can we complain about the loss of Hawaiian culture in the language of Western law? The organisers were caught between the desire to use Hawaiian words and legal concepts, many of which would be unfamiliar even to Native Hawaiians, and their desire to produce a tribunal which could be understood by the national and international audience. The organisers sought to join the two legal cultures and argued that this conjunction was necessary in order to document the wrongs of the United States. But little remains of the pre-contact legal system even among rural Native Hawaiians after nearly two centuries of legal incorporation and transformation. Both contemporary Native Hawaiians and the leaders of the Tribunal are deeply incorporated into the American legal system. The prosecutor and judges continually sought to redefine Western law even as they called on it, seeking to detach it from the nation-state context and link it to a more global notion of justice. Law acquired a fundamental pluralism. This move appears in the opening statement by the prosecutor, Glenn Morris, a Shawnee attorney at the University of Colorado at Denver and Director of the Denver chapter of AIM. He begins by linking this Tribunal to the struggles by Native Americans in the continental United States and Canada, then moves to redefining law: it is our responsibility to tell the world that justice cannot be employed without also including an indigenous vision of what justice means.

9

Use of the Sea

The seas furnish a variety of food, and they provide highways on which ships carry many of the heavy commodities that move from country to country. The seas also temper climate, making winters milder and summers cooler near their shores. The very size of the seas, which cover nearly three fourths of the area of the earth's surface with depths of up to six miles, implies their importance. Long struggles, patient work, and invention, accompanied by many lost lives, have given man a partial control of the seas; but those who know the seas respect their power.

To the early Greeks the sea was the Mediterranean, which they called "our sea" or "nurse of heroes." Odysseus, the hero of the early Greeks, travelled over most of the Mediterranean and returned to Greece; many potential heroes attempting the same exploit never returned. The Greeks first thought of the sea as an endless river of water running around the outer rim of the continents, beyond which lay the home of the gods.

As their knowledge grew, the Greeks developed the idea of the sphericity of the earth, but they still thought of the ocean as a single body of water bordering all of the land, a concept which the Romans adopted. The Greeks developed the concept of climate and thought that the ocean probably boiled at the Equator and that the land north of the Arctic Circle was too cold for any life to exist. The Greeks reasoned about the ocean, but the Phoenicians, great traders and navigators, learned about it through practical experience. Although some doubt has been cast on the legend that the Phoenician uninavigated Africa, there is no doubt at as early as 1500 B.C. the Phoenicians re sailing along the eastern coast of Africa, which they reached by way of the Red Sea. Their expeditions were obviously intended to bring back valuable trade goods to the eastern end of the Mediterranean. On, one expedition, the details of which are on record, five ships set out and brought back myrrh, ebony, gold, ivory, cinnamon, incense, eye cosmetic, monkeys, apes, dogs, and various skins as well as a number of slaves. The Phoenicians were also active along the western end of the Mediterranean, and some legends indicate that they sailed into the Atlantic and probably into the Bay of Biscay and the North Sea.

In later years the voyages of the Phoenidan stimulated the Arabs and the Portuguese to -continue explorations down the coast of Africa and eventually around it. Altogether, the desire for trade in the rare products that would bring great prices in the markets of the Mediterranean led to a very considerable knowledge of the Mediterranean and the margins of the Atlantic and Indian oceans. Later, the same desire for trade led Magellan, Columbus, and their contemporaries to travel much farther and eventually to circumnavigate the globe.

Meanwhile, scientific curiosity concerning the oceans led to many expeditions, the purpose of which was not to trade but to find out more about the seas. One of the earliest voyages of marine exploration was made by Pytheas, a Greek of Massilia (Marseilles), about 300 B.C. Though very little of the record of this voyage remains, it is known that its object was to find out how far land extended, to discover where the Arctic Circle was located, and to see the land of the midnight sun. Pytheas presumably had some sort of astronomical instruments by which he could determine latitude. In the last part of the 9th century A.D., Ottar, a Norwegian, sailed northward along the coast of Norway, around North Cape, and into the White Sea.

There is not very much known about the explorations of the Vikings in the northern Atlantic, but it is known that as early as the 3rd century there was a good deal of traffic between the seas of northwestern Europe and the Mediterranean.

Doubtless the Vikings extended their voyages as they improved their ships. Nevertheless, despite various improvements, northern sailors worked under great handicaps until the time of Columbus because they had no compass and no means of finding their position at sea (although the Greeks, of course, had had instruments centuries before). Also, since their ships were equipped with square sails, they did not find it feasible to tack against the wind.

New instruments, the magnetic compass, the sextant, and the chronometer, enabled man to find his way over the oceans; improved ships made the voyages safer and more profitable. The evolution of the modern ship was a long, slow process and improvements are still being made. Some of the earliest boats were made in Egypt of papyrus bundles, similar to the reed boats used today on Lake Titicaca, Bolivia. These were suitable for paddling, but not for rowing or sailing.

The prophet, Isaiah, referred to vessels of bulrushes on the waters of the sea (about 740 B.C.). The first Egyptian boats of wood had no keels and were simply copies of earlier reed boats. Ships of Homeric times (late Bronze Age) had well-developed keels, and the distinction between sail-driven merchant ships and oar and sail-driven warships was definite. By the middle of the 3rd century B.C., the warships had developed into two major types, heavy galleons, the father of our modern battleships, and light galleys, prototype of our destroyers. The larger galleons had two, three, and sometimes four decks of oars.

On modem ships, the only survival of oar-using warships is the round porthole. The early ships were quite seaworthy, and Hanno, the Carthagenian, coasted down the western side of Africa before 500 B.C. His claim to have sailed entirely around Africa is not accepted by all geographers.

The improvement of sailing ships and sailing techniques was a great step forward; it enabled the merchant ships to carry large cargoes with comparatively small crews. From the time of Homer to the middle of the 19th century, the sailing ships were supreme; even today a few are still in use for short voyages along the coasts of south eastern Asia, as freight ships in the South Seas, and as pleasure craft on all the oceans.

Early in the 19th century ships made of iron and powered by steam began to re place the sailing vessels, since they were more durable and dependable, and less subject to unfavourable weather. Later, steel replaced iron as the principal shipbuilding material and special types of ships were de signed for varying conditions and cargoes.

The oil tanker provides low-cost transportation for the most important fuel; the ocean going barge is used widely in coastwise transport for heavy, bulky commodities, such as coal and lumber; the large, deluxe passenger liners compete with the airplane for transoceanic travel; special types of river boats and barges are used on the inland waterways.

Of late, atomic-powered ships are coming into use, making it possible to carry large pay loads many thousands of miles without refuelling. In modern times man has, to a limited extent at least, conquered the seas. With abundant power, nuclear and otherwise, and with many ingenious craft, he moves above, on, and below the surface Of the ocean with comparative ease. Special devices enable him to penetrate the greatest depths. Icebreakers make it possible to approach shores usually blocked most of the year by ice, and an atomic submarine can move with little difficulty under the ice floes of the North Polar region.

The historic voyages of the Nautilus and the Skate under the arctic ice across the pole are of the greatest significance. Finding some difficulty in entering the Arctic Ocean by way of Bering Strait, between the ice floe on top and the shallow bottom, these submarines soon reached deep water in the Arctic Ocean. Equipped with fathometers which gave them the depth below the submarine and sonars which indicated the space between the submarine and the ice above, the ships were able to navigate in comparative safety.

Special navigation equipment was necessary since magnetic compasses are of little use in this area and, of course, it was not possible to make astronomical observations.

Both the Nautilus and the Skate encountered large schools of fish which may provide large, quantities of food some time in the future. The use of the nuclear submarines suggests the possibility of carrying freight in the future. The submarines are independent of surface storms, which do not penetrate to great depths in the ocean. Because of their ability to pass under the arctic ice it would be possible to cut nearly 5,000 miles off the present-day surface voyage from Japan or western United States to Europe.

However, these submarine voyages are subject to some hazards. One is the great undersea waves which sometimes reach heights of 250 feet or more; although they not violent, nevertheless they cause some difficulties in navigation.

An ambitious exploration of the ocean floor is contemplated in the project *Mohole*. This was conceived several years ago with the intention of drilling through the crust of the earth and into the earth's mantle. The crust of the earth is 15 to 20 miles thick under the convenient but is much thinner under the oceans. Its thickness has been measured by seismic methods and is fairly well known at different locations. It is proposed to drill through the ocean floor at some convenient point, reach the mantle rock, and find out what the content of the mantle rock is.

A great many unforeseen difficulties will probably be encountered. It should be noted that on land wells have been drilled to a depth of about five miles, but this was done with drilling rigs securely anchored on the ground. If the drilling rig is to be anchored in deep water it will be necessary to stabilize it. This can be done with the aid of computers, and such computers have been used on the drilling rigs or oil on the continental shelf.

One plan for *Mohole* is to drill off the coast of Mohole in the Hawaiian Islands. The depth at this point is about 14,000 ft, so the drill would penetrate 14,000 ft of water and about 20,000 feet of rock before encountering the mantle rock. The Mohole project may have considerable economic importance. It is possible that in penetrating the crust of the earth, where considerable heat is developed, enough energy will be provided to desalinize sea water and even to furnish additional power for other uses. Even so, man is just beginning to understand, appreciate, and utilize the ocean.

Optimists look forward to the day when plankton will be used for human food, when sea water will supply most of our minerals, and when the tides will furnish abundant power. The potential of the ocean is as vast as its surface, as unplumbed as its great deeps, and limited only by the desire and ingenuity of man. In the brief discussion below only three broad aspects of the seas will be discussed: the tides, the effect on climate, and the wealth of the seas.

Global air conditioner

Most people know that the Gulf Stream has a strong effect upon the climate of northwestern Europe, but this may be the extent of their knowledge concerning the effect of the sea on the climate of the land. Generally, the Gulf Stream is given too much credit and the sea not enough; even without the warming effect of the Gulf Stream, the sea can moderate the winter climate of the adjacent land areas.

The Gulf Stream is a river flowing on top of the heavier, saltier waters of the Atlantic Ocean. As the Gulf Stream moves along the coast of eastern North America, it does not markedly affect the climate of the adjacent lands to the west; the climate of these lands is mainly determined by the westerly offshore winds blowing over them.

The Gulf Stream is always a warm-water current, but its temperature, as well as its course, varies somewhat from year to year. Charles Brooks, a British climatologist, compared the North Atlantic to a great bathtub with a hot tap, the Gulf Stream, and two cold taps, the cold currents along eastern Greenland and the Labrador Current. Nature adjusts each of these taps from time to time, determining the temperature of the bath and affecting the weather in western Europe. Although the weather in western Europe is somewhat affected by the temperature of the North Atlantic, it is more largely conditioned by the air masses that move across the North Atlantic. Thus, the proposal of many years ago by an American anglophobe that the United States dam up the Straits of Florida and in this way cut off the Gulf Stream and freeze England was not only uncivil but unscientific: no dam could essentially divert the Gulf Stream; and even if it could, England would not freeze.

The oceans stabilize the temperatures of the earth by absorbing and storing solar energy and transferring it to the land from time to time. Although ocean waters absorb tremendous quantities of heat from the sun, they do not become hot because of their great volume; likewise, they can lose much heat and not become cold. The heat stored in the oceans is transmitted to the land by a complex combination of ocean currents and air masses.

At the contact between the ocean surface and the overlying air masses, there is continuous interaction of great importance to the climates of the earth. The air warms or cools the ocean water, depending upon the contrast between the two, but, more importantly, the ocean water acts to warm or cool the air. It takes 3,000 times as much heat to warm a given volume of water as it does to change the temperature of an equal volume of air by the same amount.

In other words, one cubic meter of water cooling through one degree centigrade will raise the temperature of 3,000 cubic meters of air by the same amount. Thus, though an ocean seems to maintain a fairly even temperature, it has the effect of lowering the temperature of the warmest season and raising the temperature of the coldest season on the adjacent lands. The effect of cold currents on adjacent lands is even more clearly marked than that of warm currents. The Humboldt Current, which moves northward along the coast of Chile, and the Benguela Current, which runs along the western coast of South Africa, combine with winds blowing from the sea over cold land to produce a climate which is almost rainless but likely to be quite foggy. Though this climate is very stable most of the year, there are important results when the temperature of the currents changes temporarily.

When, for instance, the cold water of the cold currents is replaced temporarily by warm water, there are often very severe storms on the coast of Peru and the Namib Desert of Southwest Africa. Ocean storms, to be sure, do not depend entirely upon the changes in currents. They occur frequently on western coasts in the middle latitudes. For many centuries the Bay of Biscay has been noted for its storms and its hazards to navigation and fishing, as have the western coast of North America and many other coastal areas in similar latitudes.

The Tides

The tides constitute another feature of the sea which is of profound importance to man. The difference between high and low tide is only a few inches in some places and nearly 100 feet in others. Man has learned to adapt himself to tidal changes in many ways and as benefited from the fact that the tides in most part of the world are accurately forecast. The United States Coast and Geodetic survey has an electronic computer which will automatically give the time and height of the tide on any future date at any place in the world where records have been kept.

Forecasting, however, is of limited use in he case of so-called tidal waves, or tsunamis, which are associated with earthquakes. Although such sea waves can be forecast a view hours in advance if sufficient seismographic records of the earthquakes are avail able, it is difficult to warn all people living near the shore. The tsunamis rise on the land with a tremendous volume of water and force enough to wreck ports and shipping.

Most of the people in the world live at very low altitudes, many of them only a few feet above sea level, and are, there fore, exposed to tidal waves and high tides, some of which are generated by hurricanes. For centuries men have observed the tides and marvelled at the strength of the tidal currents, wondering how this force could be turned into useful power.

The ideal condition is an embayed coastline with a narrow entrance, easily dammed and a high tidal range. At such a site on the coast of Maine an attempt was made several years go to build a tidal power station, but the project was discontinued for lack of funds. In 1966 the first unit of a tidal energy powerstation was completed on the Rance River on the coast of Brittany. One difficulty is the failure of power at slack tide, twice a day. This can be solved by using tidal power as a supplement to fuel power; in the intervals when tidal power is lacking, fuel power can be stepped up.

Wealth of the seas

The ocean has long, been a source of food, minerals, vitamins, and other materials which man can glean from its shallow margins and its greater depths. With the exception of some shallow margins which have been reserved as oyster beds or special fishing grounds, most seas are available to anyone who wishes to exploit them.

Food from the sea

Early man hunted for food in the seas at low tide, gathering the relatively immobile sea animals and plants-clams, mussels, oysters, abalones, sea snails, and a great variety of other edible materials. The great shell heaps, acres in extent and many feet in thickness, that have been found on the shores in many parts of the world attest to the long-continued use of clams, mussels, oysters and abalone as food by primitive peoples. As time went on, people began to move out over the seas and harvest a greater variety of sea food, including the large mammals—the whale, walrus, and seal—and many fishes—herring, cod, mackerel, tuna, shark, salmon, bream, sole, sardine—as well as many miscellaneous foods such as octopus, squid, and seaweed.

Most of the fish and other types of sea food come from the shadow margins of the sea—the continental shelf with depths down to 600 feet. Here the environment is most favourable for the growth of plankton, the small organisms which supply the basic food for many forms of marine life.

The continental shelf is most extensive near the coasts of northwestern Europe, eastern and southeastern Asia, and eastern North America. The dense population of these regions creates a demand for sea food so great, however, that fishing is increasing in the more distant banks of the Arctic Ocean and off the coasts of South America, Africa, and Australia.

The largest producers and consumers of fish are Japan, the United States, China, the U.S.S.R., the United Kingdom, Norway, and Canada. Of these, only Norway has a large surplus for export. The total catch for all countries is worth more than one billion dollars. The production and consumption of fish depend on a number of variable factors, which include wide expanses of the continental shelf adjacent to areas of dense population and suitable harbours for the fishing fleets. A smooth bottom is favourable since otherwise the bottom nets and trawls become snagged; in tropical waters coral reefs often prevent the use of large-scale fishing methods.

One of the most favourable locations is the contact of two ocean currents, one warm and the other cold; the mixing of these Off the coast of New England and eastern Canada from Cape Cod to Newfoundland, he continental shelf is wide and comparatively shallow; and the numerous banks, such as Grand, Georges, Sable, and St. Pierre, are heavily fished for cod, herring, rosefish, and others. The bottom is smooth, and the mingling of the warm and cool currents accompanied by fog is a favourable factor for fishing but not for shipping.

The irregular shoreline affords numerous harbours and the large urban centers nearby provide a large market for the fish. The fast, modern fishing boats are refrigerated so that the catch can be processed on board and marketed fresh. These waters are also fished by Europeans; France has bases on the islands of St. Pierre and Miquelon. The continental shelf off the coast of northwestern Europe, including the North Sea and the Baltic, is fished heavily by the people of the United Kingdom, Germany, France, the Netherlands, Norway, Sweden, the U.S.S.R., Denmark, Belgium, and Spain. Fishermen from these countries also visit the waters off Iceland and Spitzbergen.

Cod, herring, and mackerel are the most common varieties but quantities of plaice, sardines, salmon, hake, lobsters, and oysters are also taken. The herring fishermen use gill nets which are floated on the surface. The herring are small but welcomed in the diet of western Europeans because of their high fat content. The cod, bottom feeders, are taken mostly in trawls dragged along the bottom; the best cod fishing is along the northern coast of Norway. Although much of the catch of northwestern Europe is marketed fresh, quantities of herring are smoked and pickled, and much of the sardine catch is canned.

In Japan where beef production is very small, fish plays a more important role in feeding the people than in any other large country of the world. The Japanese consume about 75 pounds of fish per capita per year and the tonnage of the catch is at least twice that of the United States, the nearest rival. A part of the Japanese catch is exported, either canned, dried, or frozen, including tuna, crab, and shrimp.

A large proportion of the sea food is similar to that of other lands-herring, cod, mackerel, salmon, tuna, crab, shrimp, sardines, and flounder. Shellfish are important and, as a by-product of the oyster industry, so is the cultured pearl. But the Japanese fisheries, as a whole, are characterized by the great variety of species taken and by the different methods of fishing and preserving the catch.

Inshore fishing accounts for most of the herring, sardines, mackerel, bream, yellow tail, tuna, and flounder; in deeper waters Japanese fishing ships formerly ranged from Bering Strait to Antarctica and from Ceylon to Hawaii, taking large quantities of salmon and crab and processing them in huge floating canneries. After World War II Japanese fishing was for a time restricted to the, waters immediately adjacent to Japan.

In recent years fishing has increased on both the west and east coasts of South America, especially near Peru and Argentina; it has also increased around Tasmania, off the northwestern coast of Australia, in New Zealand waters, and on the west coast of Africa.

Man's increasing awareness of the wealth of the sea has led to increased re strictions: Iceland now forbids foreign fishing ships in their near shore waters; Peru has proclaimed a 200 mile limit and bars foreign ships from the great wealth of the nearby Humbol rent; and the U.S.S.R. guards the western Siberia.

Ocean minerals

Minerals, as well as food, are obtained from the sea. Common salt, the most important of these minerals, was the first to be exploited. Early man found that he could obtain salt by evaporating sea water in pools and shallow lagoons. Nowadays, much of the salt produced in the world is either evaporated from sea water or obtained from lagoons which have accumulated salt for thousands of years.

Blown by sea breezes, some of the salt from the sea may find its way back to the land. These small particles of salt serve as nuclei for raindrops and thus have a very important effect upon precipitation. In England as many as 25 pounds of salt per acre per year are blown in on sea breezes; Sambhar Salt Lake in northern India receives yearly about 3,000 tons of salt which have been carried in by the hot monsoon of summer from the sea 400 miles away.

The most favourable places for obtaining salt are the shallow margins of the sea where, in natural basins, sea water can be cut off and evaporated with the help of high winds, sunshine, and low rainfall. Such a place is the Rann of Cutch, an area the size of New Jersey on the western coast of India. Here the winds of the southwest monsoon carry water inland to cover the plain. Then later, as summer winds blow from the opposite direction, salt crust forms (sometimes to a thickness of several feet) in the pools which have been left all over the area. Similarly, there is 4 large bay on the eastern side of the Caspian Sea that is almost completely cut off from the main body of water. When the narrow passage connecting the bay and the sea is closed and the sea water remaining in the bay (of high salt content) is allowed to evaporate, salt deposits form and the salt is collected, Then, after the salt has been collected, additional sea water is admitted to the bay.

Petrolium

Oil-bearing sands are found underneath the shallow surface of the sea, particularly in the Gulf of Mexico, along the coast of California, and in the margins of the Gulf of Maracaibo in Venezuela. Much planning and invention were required before man could drill wells and maintain pumps in the sea, but the large seagoing rigs which are now transported from place to place represent at least a partial solution of many of the problems involved.

The discovery of oil in the sea and the techniques which have been developed to exploit it have added enormously to the world's potential supply of petroleum. Oil production, which seemed to be declining some years ago, now appears to be sufficient for the foreseeable future, perhaps for the next century. The discovery of offshore oil has raised many questions about the various ownership claims of federal governments, state governments, and private organizations.

For many centuries man has noted changes in the shoreline in part these are natural, the result of the ocean waves and currents eroding the land. But man's structures along the ocean shore have in many cases accelerated these changes. Commonly, in many parts of the world, jetties have been constructed out into the sea to improve river harbours. One effect of these structures is to change the shoreline by upsetting the delicate equilibrium of the beaches, depositing more sand, in some areas and producing serious erosion in others.

The advanced technology of modem times has resulted in serious pollution of the shallow margins of the ocean. The pollutants include sewage, both human and industrial, oil spilled from ocean tankers, sometimes as the result of wrecks, and leakage from ocean oil wells. Few beaches in the temperate belt have not been polluted by oil.

The construction of dams on the large rivers has affected the character of the marginal seas, especially with respect to salinity. A recent example is the effect the Asian High Dam on the Nile has had on the salinity of the waters in the Suez Canal.

Prior to the construction of the dam the Nile flood, reaching the Mediterranean in late summer, brought a large amount of fresh water into the Mediterranean Sea. The fresh water was conveyed eastward by currents and southward by northerly winds into the Suez Canal. It now appears that the increase in salinity will have a permanent effect on the ecology of the northern part of the Suez Canal and even of parts Of the eastern Mediterranean.

The effect of dams is even more spectacular when an inland sea is involved. The large dams constructed by the Russians on the Volga River and its tributaries have had a profound effect upon the Caspian Sea. Since much water is diverted for irrigation and industrial purposes, the sea level is much lower than formerly. The lower level is, of course, accompanied by greater salinity, which is very pleasing to the salt makers on the eastern shore but less pleasing to the fishermen.

In addition to the lowered sea level and the increasing salinity, the lower volume of water emphasizes the pollution produced by oil installations and chemical plants along the shore. Further more, the nutrient materials ordinarily carried in by the Volga are greatly reduced. Russian engineers are studying methods of restoring the level of the Caspian Sea but the difficulties are many. The most far-reaching alterations in the ocean made by man are the result of radioactive fallout.

10
Alteration of Grasslands and Forests

There are two distinct phases or aspects of man's alteration of the earth: the extensive nand the intensive. Extensive alteration is often peripheral and may come about through temporary or partial occupation, during which large portions of the natural vegetation suffer little change. Extensive changes have been brought about by grazing or browsing animals, by the selective cutting of certain forest trees, and by "patch" cultivation.

However, the time factor is important: if even partial occupation is continued long enough, there may be lasting changes. If bison, for example, invade a forest periodically and continuously for a century, browsing and trampling the young shoots so that young trees cannot grow, in the end the forest may be destroyed. Also, if shifting "patch" cultivation is practiced long enough for each plot to be cleared several times, the natural vegetation may be completely changed.

Intensive occupation, on the other hand, brings drastic changes almost immediately, especially if plow agriculture is involved. If a forest is completely cleared, with stumps removed and land plowed, the alteration is nearly complete. Even if the field is later abandoned for a time, any wild plants which appear often bear little resemblance to the original vegetation.

Fire

Man and nature have collaborated in many ways to alter grasslands and forests. Fire, both of natural origin and man-made, has been of primary significance in the process. Fire has been used by man since very ancient times, and early literature is full of records of burning and leaving fires to bum.

One of the oldest historical records, the Periplus of Hanno, describes fires in Africa: "... by day we saw nothing but woods but by night we saw many fires burning... going along four days, we saw by night the land full of flames and in the midst was a lofty fire, greater than the rest and seemed to touch the stars." There is also evidence of man's use of fire in prehistorical times.

Deposits in the caves occupied by peking man indicate that he used fire more than 200,000 years ago, even though he may not have been able to m it himself. The use of fire is one of the m criteria which distinguish man other primates. It is reasonably certain that fires caused by lightning, volcanic activity, and other natural phenomena occurred long be ore the advent of man. It remained for man to adopt nature's fire, learn to make fire himself, and use it for his own purposes.

Ancient man did not attempt to put out fires set by lightning, nor did he bother to put out his own campfires. It was often an advantage to him to have fires spread, since a closed forest does not have large numbers.

The bison occupied the forest, but they grazed mostly in the large clearings which had originally been produced by the Indians' burning of the forest and brush in order to improve game conditions. Since the pioneers had few domestic animals and lacked sufficient domestic cattle for meat, they killed the bison in large numbers and, in a matter of a few years, drove them out of the grasslands on the forest margins.

Modification of the grasslands

Since the natural grasses are valuable to man only after they have been converted by animals into high-protein foods, the grasslands were first used by hunting peoples; later the grasslands were used by agricultural and pastoral peoples who did some hunting in the margins of the agricultural areas.

In Northern America the first hunters to record their contact with bison were pioneers who crossed the Appalachians moved into Kentucky and Tennessee. The bison occupied the forest, but they grazed mostly in the large clearings which had originally been produced by the Indians' burning of the forest and brush in order to improve game conditions. Since the pioneers had few domestic cattle for meat, they killed the bison in large numbers and, in a matter of a few years, drove them out of the grasslands on the forest margins.

As the settlers moved west, they encountered two distinct types of grassland: the humid prairie and the semiarid short-grass country. The prairie was the first of the grasslands to be altered on a large scale, as farms were established in such states as Indiana, Illinois, and Iowa.

Although the tough sod of the prairie was initially difficult to plow, such plowing was easier than the clearing of forest land. The settlers were surprised to find that when a field was plowed and then abandoned, it had a succeeding growth which consisted largely of shrubs and small trees rather than grass.

Defined as a grassland in a forest climate, the prairie supported forest trees along its stream valleys. Although the principal result of the occupation of the prairie was the replacement of grass with cropland, a further result was the conversion of parts of the prairie into woodland.

Beyond the prairie were the Great Plains, the short-grass country, where buffalo grazed on the grama grass which grew in bunches with bare ground in between. That this region was often called the Great American Desert is attributable to the fact that some of the early surveys of the area were made in drought years; in humid years the Great Plains did not resemble a desert.

A century or more ago many people thought of all treeless regions as deserts and so called the Great Plains a desert; others thought the presence of grass indicated that it was a more humid region. Those who opposed western settlement for political reasons called the region a desert; those who stood to gain by settlement were sure this region would blossom as soon as settlers moved in.

Modification of the forests

The modification of the forests of the earth has varied from place to place and from time to time. It will be possible here only to note some of the changes brought about in the mid-latitude and northern coniferous and deciduous forests. In many cases man's exploitation of the forest began with the use of the forest animals.

An example of this initial exploitation is the early fur trapping in North America, particularly in northeastern United States and Canada. In the process of trapping beaver, the principal source of furs, the fur trappers made only slight modifications in the forest. Undoubtedly they occasionally set fires, but their chief use of the trees was for birch bark for their canoes and fuel for their campfires. Some modifications were made along the trails, particularly the portage trails between the lakes and streams, but these alterations were minor.

Large-scale alteration of the forest began with the actual use of timber in some quantity. At first the effect of this use was decidedly peripheral. In New England they began to use the forest for ship masts, spars, and shipbuilding timbers in general. The spruce was used for masts and spars, and the birch, particularly the knees of the birch, was used for ship timbers. There was a more complete exploitation of the forest land as the population increased and lumber was used in large quantities for the construction of houses and furniture.

The New England forest was a mixture of coniferous and deciduous trees. Some deciduous trees later considered valuable were not used much in the early days; one of the trees neglected in the early days was the sugar maple, which later came to be used both for its syrup and as a furniture wood.

In the coniferous forests of northeastern United States and the adjacent parts of Canada at the time of active lumbering operations more than a century ago, the principal tree was the white pine. The white pine occurred in nearly pure stands and could be cut cheaply in large quantities and readily sawed into lumber and other products.

The widespread logging of this tree, however, resulted in permanent change in the vegetation because the pine did not reseed itself very well when large areas were cut. The successive vegetation was likely to be brush, consisting of quaking aspen, scrub oak, and various shrubs.

Later, when large quantities of spruce were cut for pulp in the more northern forests, other trees growing along with the spruce (such as the balsam) were left standing. This practice resulted in a permanent change in the forest but not in its complete destruction.

Clearing of the forest for farms

Clearing of the forest was selective since certain types of forest were regarded indicative of certain qualities in the land. For example, the pioneers crossing Appalachians from Virginia and the brought with them a general notion that the most fertile land was indicated presence of hardwood trees, particularly the black walnut and the white oak. As a result of this belief, the large hardwoods were cut in many places and were destroyed by burning, there being little demand for the lumber in the particular localities.

One common method of beginning a clearing was to girdle the trees by cutting through their bark a few feet above the ground, causing their death. Crops could be cultivated among the dead trees since they would have no leaves and, therefore, would cast little shade. Such a clearing is called a deadening.

In recent years the rapid destruction of the forests has been offset by tree farming at is designed to bring about a sustained yield. Tree farming has been going on for several generations of trees in western Europe and Japan, but the practice is just getting under way in the United States.

Tree farming may be carried on in various ways: it may include planting pruning thinning and harvesting; it may often be only carefully planned cutting, in conjunction with fire, insect, and rodent control; in other instances it may include seeding transplanting, and even irrigating.

The great increase in tree farming in recent years has come none too soon. About half of the world's forests have been cut over; some of the land is now in farms and some is being reforested, but large areas some is being reforested, but large areas consist of cut-over waste land in which little or no regrowth occurs. It is estimated that there are three billion acres of potential forest land in the world, including tropical forests and inaccessible forests in isolated regions.

Tropical forests are so extensive that they seem to offer the greatest resource of usable trees, but many authorities insist that the broadleaf forests of the tropical lands contain only a small percentage of usable trees and that these are scattered through the tropical swamps and uplands, making logging very expensive except near navigable water. In this century at least, it may be more feasible to grow timber by careful management in the midlatitudes than to harvest it in the tropics.

Consequences of man's alteration of the land

Man's occupation of the land and the consequent modification or destruction of the natural grasslands and forests have had far-reaching effects, some of which are not completely understood.

Man's effect on climate, for example, is a controversial subject. Debate centers around such questions as whether or not the destruction of forests has any measurable effect upon climate. It has been contended that forests increase rainfall and that the destruction of forests decreases rainfall. This contention is difficult to prove, however, and evidence indicates that if rainfall is increased by forests, the amount of increase is probably slight. This question was argued during the middle 1930's with reference to the Dust Bowl in the Great Plains.

It was proposed at that time to construct a shelter belt of trees, from the Canadian border to the Mexican border, which would, according to its proponents, increase the rainfall, decrease the dust storms, and have many other beneficial effects.

It is true that in forests temperatures vary somewhat from those of adjacent grasslands presumably receiving the same amount of energy. However, in areas of thin soil, where bed is at or near the surface, temperatures during the warmest part of the day are notably higher than they are in forests. The result is increased evaporation, a drier environment, and a quite different vegetation pattern.

Another effect of vegetational changes concerns the rate of runoff of streams. It is generally conceded that the destruction of natural vegetation, particularly the forests, tends to increase flood hazards. However, it should be pointed out that destructive floods occurred long before any great changes in the vegetation cover were made.

For example, the greatest recorded flood in the Willamette Valley of Oregon occurred in 1861 before any substantial settlement had been made in the watershed of the river or any clearing of the forests had been made on-a large scale. The flood was probably caused by the combination of unusually heavy rainfall and rapid melting of a heavy snow cover in the adjacent mountains.

Man has changed the face of the earl in many ways. Although his changes have affected much more than the grasslands and forests, many of them have been unplanned and even unintentional. Many people consider that it is possible to alter the landscape wisely only in those areas with surplus raw materials or some other kind of surplus wealth. However, even in the richest of lands there is little evidence that man is inclined to alter the landscape wisely until most of the original natural resource has been destroyed.

11
Communications

The effect of communication is far-reaching and continuous. In general, communication makes people more alike. Nowadays, as people are exposed to ideas from many lands, they tend to adopt some of the ideas of other people, language barriers break down, and ways of life are modified. Conversely, communication also serves to intensify differences between regions: the exchange of goods may enable one region to specialize in a certain kind of production and thus become different from the adjoining region.

In the early days, two agricultural regions fairly close to each other might be growing the same crops and producing the same staple foods, textiles, and clothing. With the establishment of transportation between these regions, however, each would be able to specialize to a certain degree in some products and to give up the production of other things. Thus, one region might specialize in wheat while another specialized in flax, the one producing a staple food and the other producing a staple fiber. Of course, there are many barriers to transportation which tend to limit its effect: transportation costs, tariffs, and embargoes, for example. The first roads used by man were animal tracks.

The trails of grazing and browsing animals might be long, straight, or crooked, but they usually followed the easiest route across the terrain, whether it be forest, brush, or grassland. At first, man moved along these trails as he hunted the animals, but later the animal trails became permanent human trails which eventually became modern roads. Many of our early roads were so obviously animal trails that they were sometimes referred to as cow paths. In Africa elephant paths provide easy routes through forests which otherwise would be very difficult to traverse. Antelope tracks serve as trails in other areas, and in the middle part of North America buffalo trails were used by the early travellers and hunters. Even small animals can, by continuous use, make a trail large enough for a man to travel. Man used the early trails and roads to get his prime needs—food, wafer, and fuel.

The earliest roads probably led from such primitive shelters as caves to streams or shores where water and fish could be obtained and into woods and prairies where game could be hunted. In modern times the very diverse media of transportation have brought great changes in man's relationship to the natural environment. Man's concept of a mountain range changes when a new road or a new kind of transportation is involved. The kind of transportation which is best suited to a region, in terms of cost, speed, and safety, is gradually developed and becomes predominant there.

Transportation employs one fifth of the world's manpower and energy. In primitive societies the methods of moving people and goods are laborious, with the result that goods do not move very far and the volume of goods transported is small; muscle power, human and animal, is the chief moving force. In an industrial country about one person out of five is engaged directly in transportation or in the servicing and selling of goods that are used in transportation. Consider, for example, the large number of people engaged in servicing and operating automobiles, trucks, trains, and airplanes. Urge amounts of mechanical energy, derived largely from coat and petroleum, are also used in transportation.

Variety of carriers

The people, goods, and ideas of the world are transported from place to place by a variety of carriers: there are still human porters carrying goods in parts of Africa and Asia; there are goods being moved by wheelbarrow, cart, automobile, airplane, train, truck, jeep, pipeline, and various kinds of pack animals. On water a great variety of craft are used from sampan to ocean tanker. Each type of carrier can be evaluated in terms of the cost of transporting one ton for one mile, but every carrier has qualities which cannot be shown by a ton-mile evaluation.

The ocean liner, for example, can carry goods to the land, but no farther; the truck can carry goods to particular localities, from door to door, while the railroad can deliver only to a station; human carriers and pack animals can penetrate areas that do not have roads or even adequate trails. Thus, a comparison of carriers that is based solely on ton-mile costs is not entirely valid. Limited in range and speed, human porters today can transport goods for about one dollar per ton-mile; the cost for pack animals in most parts of the world is about the same.

Two horses and a wagon, formerly a very important mode of transport, can move goods at a cost of from 10 to 25 cents per ton-mile. The motor truck, some form of which is now used in most parts of the world, can transport goods at a cost of about 10 cents per ton-mile. The main advantage of the truck lies in its ability to travel over all roads and reach remote localities. It can carry its load from an original source, like a wheat field, to a processing point, like a flour mill; it can carry coal from a mine to its place of consumption. Railroads, which have played such an important part in world transport in the last century, can move goods at a cost of from one to five cents per ton-mile. Naturally, the cost will vary with the nature of the material carried; coal, oil, wheat, and other bulk goods can be carried more cheaply than packaged goods, which have to have special handling.

The ocean freighter can carry goods more cheaply per ton-mile than any other form of transportation, the cost running from two tenths of one cent to about one cent per ton-mile. Whenever long hauls are possible and speed is not a factor, the ocean freighter is a preferred means of transportation. Today, the airplane carries people and goods of all kinds to remote localities and offers the advantages of speed and long range.

The cost per ton-mile varies widely, but it averages about 25 cents for air freight. The pipeline is only useful in transporting gases and fluids, such as oil and gasoline, but it can carry these materials very efficiently and at low cost. The cost of pipeline transportation is slightly higher than that of transportation by tanker if the hauls are comparatively long.

Early roads and trails

The first routes of any length were mere trails developed in order to make widely available some of the valuable goods which could be obtained in only a few localities. Such valuable goods included amber, lapis lazuli, tin, and silk. Amber, a fossilized gum excreted by coniferous trees, was produced in ancient times on the shores of the Baltic Sea. From the Baltic Sea the amber, which was considered to be a precious stone, was transported across Europe to the Mediterranean along trails which are indicated today by the occurrence of small pieces of amber in areas where it was not produced.

The earliest amber road led from the vicinity of Danzig (Gdansk) on the Baltic southward through what is now Poland and Austria to the Adriatic Sea in the vicinity of Trieste. Later, other amber roads were developed which ran south from Denmark by way of the Elbe River to Czechoslovakia and reached the Adriatic in the vicinity of Venice. These roads, together with their several branches or alternate routes, were the first roads to be used in Europe.

The trade in amber, which was known in the Mediterranean in the time of Homer, began about 2000 B.C. and continued, with some interruptions, for 20 centuries. Communication across Asia in the early days was limited by terrain and climate to two routes. One of these routes began just to the south of the Black Sea, continued eastward to the south of the Caspian Sea, and ran through Turkestan and on into central Asia. This is essentially the route followed by Marco Polo many centuries after the first roads were built.

A more southerly route led through Turkey, Mesopotamia, and Persia and provided for the transportation of silk and a few other very valuable goods. As early as 3000 B.C. lapis lazuli was a very important trade item upon which routes were based. (This semiprecious gem has been found along the early routes which led across Afghanistan, Persia, and Arabia.) Blue in colour and very scarce, lapis lazuli was used both as a gem and as a pigment and was especially valued by the Persians and people in neighbouring countries.

In western Europe a number of the early routes were developed by the traffic in tin. Tin was in great demand in the Mediterranean, particularly in the Roman Empire, where it was used as an alloy with copper to form bronze. Perhaps the first movement of tin was from Cornwall, a source of tin in southwestern England, by boat to the Baltic and thence overland to the Mediterranean along the amber road. The later tin routes which were developed across France and Italy to the Mediterranean required that the tin be ferried across the English Channel from the vicinity of Dover to one of the French ports, usually Boulogne, and carried overland to Marseilles on the Mediterranean by pack horses. So that it could be easily packed on horses, the tin was cast into large ingots, shaped, and notched; an average load consisted of two ingots weighing 130 pounds each. Although such commodities as amber, lapis lazuli, silk, and tin were also carried by sea, the land traffic had the effect of establishing through routes in regions which had previously had only local trails.

Roman roads

The first great Roman highway, the 162 mile-long Appian Way, was begun in 312 B.C. It was at first merely gravelled, but two decades later paving was added. Two typical Roman roads are shown in Figure 22.2 in cross section. The cost of transporting goods on these roads was great, and it is estimated that the value of a load would double in the first 100 miles. From about 500 B.C., special messengers on horseback could travel 90 miles per day on roads and levelled tracks. This speed was average until after the early 1800's. A century after the starting of the Appian Way, the Flaminian Way, linking Rome with the Po Valley, was built. There after, these were continuously added to, and new roads built.

The Roman Empire was the gift of its roads, just as Egypt was said to be the gift of the Nile. Only after A.D. 200 did the pace of road building in the Empire slacken. The first great continental system of well-built roads was constructed by the Romans. This system, originating in Rome, extended as far as Scotland to the northwest and beyond Jerusalem to the east. The longest road in the system was nearly 4,000 miles long, and the total length of the Roman road system at its maximum development was more than 50,000 miles. The quality of the roads varied widely from place to place; some sections were paved, either with slabs or cobbles, and some were not. Although the Roman roads are sometimes characterized as solely military, the fact is that they were usually constructed after a country had been conquered in order to maintain traffic for administrative purposes and to allow the movement of peoples and goods. Soldiers and common people moved along the roads by foot at the rate of about 20 miles per day; the rate of movement for the post and officials on horseback was about 40 to 50 miles, a day. Chariots, or stages, were used for special purposes and Caesar, by using a series of chariots, once travelled 100 miles in a day. Using three relays of chariots, the emperor Tiberius once sped from Germany to Lyon in France at the rate of 200 miles a day.

Inca roads

One of the most remarkable of the ancient roads was the road built by the Incas of Peru. Stretching from Quito, Ecuador, into central Chile, this Toad was about 20 feet wide and paved by blocks, even though the Incas had no vehicular traffic. Used by foot travellers and pack animals (such as the llama), the road climbed mountain ranges 14,000 or 15,000 feet high by easy grades and switchbacks; it crossed canyons thousands of feet deep, bridged roaring torrents, and, in some places, reached into the desert. Humboldt described the road as having a deep understructure and being, paved with hewn blocks of black porphyry. According to Humboldt, none of the Roman roads in Italy, southern France, or Spain was more imposing than this road of the Incas.

Road materials

The main features of a road are its grade, its curves, and, above all, its materials of construction. For thousands of years man has attempted to place a surface material on his roads and trails that will make them easily travelled in all weather conditions. Since early man used whatever material was close at hand, the first material to be used was soil, which was followed by broken rock of various kinds, including gravel usually obtained from stream beds or glacial outwash. Many kinds of crushed rock, including limestone and basalt, are used in the construction of modern roads. Of the many kinds, some of which are much better than others for road construction, limestone is particularly good because the limestone dust produced by wear forms a cementing material.

The famous pikes of a century ago were constructed mostly of limestone and served their purpose very well. As they hardened, they became fairly smooth, all-year roadways which provided very satisfactorily for the slow-speed wheel traffic of wagons and buggies. Eventually, the introduction of high-speed automotive traffic rendered these roads inadequate. Roads are also commonly surfaced with blocks of various kinds, including bricks, cobblestones, and wood. Some cobblestone roads have endured for centuries, but cobblestones do not usually make a smooth road. Wood blocks have been used in many cases, but they are subject to expansion and contraction with changing moisture conditions. (Usually, wood blocks are impregnated with asphalt, creosote, or some other material to make them last longer.) Bricks have been used for roadways, particularly in eastern United States, with fair success.

Modern roads

In modern times most roads are constructed of concrete, asphalt, or a combination of the two. Both of these materials can be applied in even layers of uniform character to produce roads which endure well under ordinary conditions. Concrete is more expensive and a little more difficult to apply than asphalt, but some modern roads have been successfully built of a mixture of asphalt and concrete. In constructing an asphalt road, it is necessary to provide a thick base of stone, gravel, or other material and also to provide for subsurface drainage. These principles were first discovered by John L. McAdam in England, who wrote:

> That it is the native soil which really supports the weight of traffic; that while it is preserved in a dry state, it will carry any weight without sinking and that it does in fact carry the road and carriages also; that this native soil must be previously made quite dry and a covering impenetrable to rain must then be placed over it in that dry state; that the thickness of the road should only be regulated the quantity of material necessary to form such impervious covering and never by any reference to its own power of carrying weight.
>
> As no artificial road can be made so good as the natural soil in a *dry state*, it is necessary to preserve this state. The first operation should be the reverse of digging a trench. The road, should not be sunk below, but raised above, the adjacent ground; that there be a sufficient fall to take off the water, so that it should be some inches below the level of the ground upon which the road is, either by making the drains to lower ground or, if that be not practicable from the nature of the country, then the soil upon which the road is to be laid, must be raised some inches above the level of the water.
>
> Having secured the soil from *under* water, the road-maker is next to secure it from rain water, by a solid road, of clean, dry stone, or flint, so selected, prepared and laid, as to be impervious to water; and this cannot be effected, unless the greatest care be taken, that no earth, clay, chalk, or other matter, that will hold water, be mixed with the broken stone; which must be so laid, as to unite by its own angles into a firm, compact, impenetrable body.

> The thickness of such a road is immaterial; as to its strength for carrying weight, this object is obtained by a dry surface, over which the road is to be placed as a covering, to preserve it in that state: experience having shewn, that if water passes through the road and fills the native soil, the road, whatever its thickness, loses its support and goes to pieces.'

Although many other road materials (including rubber, cottonseed, and iron or steel plates) have been used experimentally in small quantities, they have been proved too expensive for general use despite their advantages. The culmination of modern road construction lies in the modern freeways and toll roads, with their multiple lanes, limited accesses, and elaborate cloverleaf intersections and exchanges. These roads, constructed with easy curves and gentle grades, enable traffic to move at high speed through congested areas. Freeways serve to channel an even greater percentage of the passenger traffic into private automobiles.

Car, Bus, and Truck

The motor car, bus, and motor truck revolutionized the use of roads all over the world by providing more rapid transport for passengers and freight. A motor car carrying five or six people At a cost of roughly 10 cents a mile can compete very successfully with railroads and airlines on a cost basis. Although the airline is more rapid, the lower cost of the motor car, plus its added convenience, makes it the largest carrier of passengers in many parts of the world.

The bus is even more economical than the automobile and almost as convenient. In the carrying of freight, the motor truck has many advantages over other means of transportation. For one thing, it can carry small units efficiently. Also, it has reasonable speed and lower rates for short haul door-to-door transportation; neither the airlines nor the railroads can handle door-to-door traffic. In addition, the schedules of trucks are more flexible; their services are always available since they are on the highways day and night. For many types of freight, truck transportation means less packing, and less damage in transit. On the other hand, there are some drawbacks to truck transportation: for example, trucking charges are relatively high and by no means uniform; many truckers have only limited liability; and there is a great deal of irregularity in the scheduling and regulation of truck traffic. For very long hauls of large bulk shipments, rail transportation is cheaper. A combination of rail and truck transportation, called "piggy-back," in which the trailer portions of larger trucks are loaded on flat cars, is widely used. This method combines the low rail rate for long distances with the flexibility of trucks for door-to-door delivery.

Railways

Nearly every country today depends more or less upon railroads to transport heavy goods over relatively long land distances. In the United States railroad construction began in the early part of the 19th century. At the beginning of the 19th century the chief Atlantic ports—Boston, New York City, Philadelphia, Baltimore, Wilmington, Charleston, and Savannah—were all about the same size, and a contemporary observer might have hesitated to say which of these cities would ultimately be the largest. Many people thought that Philadelphia might become the largest because at the time it was somewhat larger than any of the others. Before the construction of railroads there had been a widespread interest in canal construction; therefore, others thought that New York might become the largest since the success of the Erie Canal had given it an advantage over the other cities and caused it to spurt ahead in population growth. Not all the canals were as successful as the Erie.

The Chesapeake and Ohio Canal, for example, which was planned to connect the Potomac River with the Ohio River, was not successful, never having been entirely established as a water route (a short rail connection was necessary over the crest of the Appalachian Mountains). Once rail construction was undertaken, it proceeded very rapidly and involved most of the Atlantic port cities. The Pennsylvania Railroad was completed from Philadelphia to Harrisburg in 1852; the Boston and Albany was completed in 1841, shortly before the Boston and Maine.

The big problem in rail construction in the United States; as in many other parts of the world, lay in mountain barriers. Once the Appalachian barrier was crossed and the interior lowland was opened, railroad expansion was very rapid indeed. Today the railroads of the United States reach all parts of the country, the densest network being in the eastern half. The New England group of railroads serves the northeastern part of the country and carries large quantities of Canadian wheat to such ports as Boston and Portland, Maine, when the St. Lawrence Seaway is closed by ice.

The Trunk Lines connect New York City and Chicago and carry large quantities of freight, including grain and coal, as well as passengers, mail, and express. Of the Trunk Lines that cross the Appalachians, the New York Central has the easiest route—up the Hudson River, across the Mohawk Valley to the Lake Ontario Plain, and thence close to the Great Lakes all the way to Chicago. The Coal Roads, another group of railroads, are designed primarily to get the coal out of the Appalachian coal fields. The Granger Lines connect the upper Midwest area, including Chicago, and the northwestern part of the United States; these lines include the Northern Pacific Railway and the Milwaukee Road.

The Southern Roads generally run north and south, connecting Chicago to the Gulf ports of Houston, New Orleans, and Mobile; the Illinois Central Railroad and the Southern Railway are examples of these roads. The Southwestern Roads connect Chicago with what might be called the Old Southwest, mainly Texas and New Mexico; an example of these is the Missouri-Kansas-Texas Line.

Finally, there are the Transcontinental Lines, which mostly connect Chicago with the ports on the Pacific. (Obviously, the Granger Lines could be classified as transcontinental in this sense.) Only one of the Transcontinental Lines, the Southern Pacific, does not come into Chicago. Whereas the Southern Pacific is a truly transcontinental line, in that it reaches from Pacific Coast ports to Gulf ports (specifically New Orleans), the other Transcontinental Lines reach only to Chicago and connect there with the Trunk Lines to the Atlantic.

Railroad maps show the density of the railroad network in the different parts of the country. The very open network of railways west of the 100th meridian is related to the barriers of mountains and the equally important barriers of dry, unproductive areas in which little or no traffic is to be picked up. In the eastern half of the country, which has a generally dense network of railways, there are some areas with few railroads; examples of such areas are the hilly or mountainous sections of the Ozarks and the Appalachians and somewhat remote districts like northern Maine and northern Minnesota.

The Canadian railroads, concentrated in the southern portions of Canada, are closely connected with those of the United States. Two Transcontinental Lines, the Canadian Pacific Railway and the Canadian National, connect the ports on the St. Lawrence Gulf with the west at Vancouver and Prince Rupert in British Columbia. In

Mexico the rail lines are widely spaced. Railroads serve as connections with the United States and carry a fair share of the Mexican freight, but Mexico depends much more than many countries upon truck transportation.

In South America railroad building was started very late and has not reached a high stage of development in most parts of the continent. Only in small parts of southeastern Brazil, eastern Argentina, and Middle Chile is there an adequate railroad net. It has been said that South America jumped from the age of human and pack-animal transportation into the age of air transport without going through the railroad stage of development. This is at least partially true, but at the present time some additional railroad lines are being constructed in South America, notably those connecting southern Brazil with Bolivia, northern Argentina, and Chile.

Only the southern and northwestern parts of Africa have fairly adequate rail connections. Some rail connections have been established between especially productive areas in the interior, such as the Katanga copper district, and the coast. It is possible to cross Africa by rail from east to west but not from north to south.

In western Europe, particularly Germany, France, Great Britain, Switzerland, and the Low Countries, a very dense network of railways exists. Since rail transportation came to western Europe after the cities were well established and a very elaborate system of roads and canals was already in use, the railroads followed the already established lines of communication. It should be noted that some European railroad stations are to be found on the periphery of a city rather than in the center (unlike many American railroad stations). The railroads of Europe face difficulties of terrain. Even in the Paris Basin, which is considered to be a fairly level region, it has been necessary to construct some tunnels in order to make easier grades. In general, tunnels are more numerous in Europe than in other parts of the world; many European tunnels, such as the spiral tunnel in the Swiss Alps, have been built not to go through mountains but to make gentler grades over them.

The rail network of the U.S.S.R. is generally open, but it plays a very important part in the economy of that country. In recent years the Russians probably have built more new rail lines than any other country. They have especially concentrated on the routes in central Asiatic U.S.S.R. and on the extension and improvement of the Trans-Siberian Railway; they have also given attention to the Turk-Sib Railway, which connects the Caspian Sea with the Trans-Siberian Railway by way of south-central Asiatic U.S.S.R. Many parts of the Soviet Union, particularly northern Siberia, have no railroads at all and depend mainly upon air transportation.

Airways

This is not merely the Air Age, it is the jet Age. Throughout the world airplanes have taken over a good share of the world's transportation. The airplane has, first of all, a great advantage in speed. Even the slowest airplane in current use is likely to be faster than any other type of carrier, and the fastest commercial planes now approach the speed of the earth's rotation (while military and experimental planes, of course, far exceed this speed). Another advantage of the airplane is its ability to travel in direct routes; unless detoured by local and temporary severe weather conditions, planes can fly great circle routes for long flights and take the shortest routes over mountains, deserts, and water barriers.

In many places the airplane has competed so successfully with other carriers that they have been eliminated entirely. For example, airlines from the Pacific Northwest to Alaska now carry all the passengers, mail, and express; the regular passenger ships formerly in service have disappeared completely, and only tankers, freighters, and the cruise ships with summer tourists make the voyage. In Hawaii planes have long since taken over most of the passenger traffic from the mainland and have taken over all the interisland traffic formerly carried by coastwise steamers and motor ships. One can now travel from Oahu to the Island of Hawaii by plane in an hour, whereas the trip formerly required an overnight voyage which might be very rough. That the competition of planes is not limited to areas of water transportation or to areas without rail and inland transportation is illustrated by the fact that planes have largely replaced the boats on the Magdalena River in Colombia which, in conjunction with short-line railroads, carried passengers and freight from Barranquilla on the northern coast to the capital at Bogota. Airplanes can function in places where there are no roads or other carrier routes and can even operate without specially prepared airfields or airports.

The airplane is versatile partly because it can be constructed in various-sizes and out-fitted with various kinds of equipment. In Canada and Alaska the floatplane, equipped with skis in the winter, effectively provides transportation in the vast northern areas. At Anchorage, Alaska, two small lakes form the airport for hundreds of floatplanes, many of them privately owned. Since Anchorage has limited contact with its hinterland by rail and road, the planes, served by the numerous lakes which can accommodate floatplanes in summer and ski-equipped planes in winter, go a long way toward solving the transportation problem. The planes are inactive only during the period between the freezing and thawing of the lakes when neither floats nor skis can be used. There are planes larger than floatplanes also being used in the area; unlike floatplanes, these planes require established air routes with airfields, beacons, and other adjuncts to navigation. Weather reports are, of course, of great importance to all types of air traffic.

The principal airways of the world are concentrated in North America and western Europe and are interconnected by many transatlantic lines. The main world route might be described as originating in New York City and connecting western Europe southern Europe, India, southeastern Asia, and Australia. The heavy traffic on this air route is carried by a number of airlines from various countries. The transoceanic airlines carry a volume of tourist traffic that is far greater than that carried by passenger ships.

In North America the principal air routes run east and west, much like the main railroad routes. Some less important air routes run north and south, and a number of fines have routes going both east and west and north and south. Air traffic is heaviest in the northeastern part of the United States and the adjacent parts of Canada—areas of intensive industrial development. The major cities are all connected by a number of airlines; such cities as New York City, Chicago, San Francisco, and Los Angeles both initiate much traffic and handle connecting air traffic. At Chicago one passenger plane takes off almost every minute from one or the other of the two major airports. New York City with its several airports has an even greater density of flights. Some cities, however, do not function in any large measure as exchange points; not many passengers en route to other cities actually deplane in such cities as Philadelphia, San Antonio, Albuquerque, El Paso, Reno, and Spokane.

In South America the principal air routes are peripheral to the continent. One route from the eastern coast of North America leads almost due south to strike the western coast of South America. Other routes lead down the eastern coast of South America and cut directly across the Amazon River without following the line of the coast in great detail. Although short connecting routes lead into the interior at many points, there is not a dense network of airlines in South America.

Western Europe has the world's heaviest concentration of airlines, and cities like London, Paris, and Berlin have very heavy air traffic. From these cities the traffic radiates in all directions: across the Atlantic, down the western coast of Africa and thence to southern South America, eastward into eastern Europe, and southeastward into Asia. Even though many of the flights in western Europe are short, the plane competes very successfully with rail and highway transportation. Ground traffic is congested and slow, especially where bodies of water or high mountains are encountered.

In Africa the principal routes along the western coast and north and south along the Nile into the Cape region (the old Cape to Cairo route on which a rail line was planned but not completed). Since Africa has very few railroads and highways in the interior, planes are often used for short, local trips. It is much easier, quicker, and cheaper to fly from the western coast to the Katanga copper area than to travel there by boat and rail.

In the U.S.S.R. the airplane has been very important to the remote sections of the country. Although the percentage of traffic devoted to passengers has not reached that of other countries, the plane carries much of the passenger, mail, and light express traffic in central Asiatic U.S.S.R. and Siberia. Countries like Japan, India, and Australia, remote from the industrial areas of Europe and North America, are well served by airlines. Japan and southeastern Asia are reached by airlines across the Pacific, one main airline flying via Hawaii and another travelling the shorter, great circle route via Alaska and the Aleutian Islands.

Growth of air transport

Altogether, the growth of air transportation during the last 40 years has been phenomenal; and continued growth is evidenced by the development of larger planes with greater payload efficiency and the increasing amounts of freight and express (as well as more passengers and mail) being handled by the airlines. A part of the growth of airlines has been the result of various government subsidies, especially mail subsidies. It has been common governmental practice to pay airlines sums of money for carrying the mail which are far in excess of the postage received. It is considered good practice in many countries to encourage commercial airlines, as part of the national defence program, to increase the number of planes and to improve the airports and other facilities. In many countries the airlines are government owned and operated.

Formerly, only high-priority materials and goods could be carried efficiently by airlines, but it is apparent that more and more commodities will be transported by air as the carrying efficiency of planes increases. Already coffee, livestock, and many other commodities are being carried by planes where other carriers are not readily available of her because of the nature of the roads or railroads, the ground traffic is very heavy.

The development of world airways has made profound changes in the economy of some parts of the world. It has, for example, made it possible to transport oil drilling equipment and other pieces of heavy machinery, as well as the men and supplies to operate them, into out-of-the-way places. Huge gold dredges weighing hundreds of tons have been dismantled and flown into remote districts. The airplane competes successfully with other carriers in the transport of high-value commodities over long distances, especially when speed is a factor. For short hauls, 100 miles or less, the truck and motor car are more efficient and more flexible as to schedules and terminals. Even for short hauls, however, the helicopter is used more and more. And as larger and larger aircraft are built, the cost per ton-mile will be reduced and airlines will be able to transport an even greater share of the world's goods.

Other carriers

In addition to roads, railways, airlines, and ocean carriers, which carry a large part of the world's traffic, such other carriers as inland waterways, pipelines, and power lines have a share in the total transportation picture. Pipelines carry oil and natural gas from the producing fields to the markets or to ports where tanker service is available. The distribution of natural gas is almost completely dependent upon pipelines, whereas oil can also be transported by tanker, rail, and truck. Pipelines also carry water to cities and areas of irrigation.

The use of power lines to transmit energy for hundreds of miles has had a profound effect on industry, serving in many cases to decentralize it. Paralleling the expansion of transportation, the easier transmission of ideas by means of the printed word, telephone, telegraph, radio, television, and the spoken words of millions of world travellers brings all parts of the inhabited earth closely together; no one part of the world can now be independent of the others. This interdependence is a new phenomenon, the effects of which are just beginning to become apparent.

12
Multiculturalism and Individualism

The term 'multiculturalism' covers a number of current political trends in North America and elsewhere which, although they are quite different in their aims and ideological content, share a positive evaluation of cultural traditions and, particularly, the cultural or ethnic identities of minorities. Multiculturalism is evident in literature and the arts as well as in politics, and it seeks to revalorise the artistic and intellectual contributions of hitherto silent minorities as well as supporting their quest for equity in greater society. Related to the critical Hegelianism of the early Frankfurt School, to feminist critiques of epistemology and to postmodernist trends inspired directly or indirectly by Derrida, multiculturalist thought is often accused of inspiring nihilism since it seems to relativise absolute value judgements.

This chapter is restricted to a discussion of one particular political aspect of multiculturalism, and investigates under which circumstances multiculturalist ideas may be at odds with individual human rights (as depicted in the original UN charter). As a consequence, it is necessary to review the concept of 'culture' invoked in multiculturalist thought. This conceptual discussion (which has practical ramifications) forms the head and tail of the article, the main body of which is devoted to a critical presentation of multiculturalist practices and debates in Mauritius, which is used here as an exemplar of multiculturalist dilemmas and opportunities. For many years, it was commonplace within post-evolutionist comparative cultural research—cultural and social anthropology -to assume that cultures were generally sharply delineated and distinct, relatively homogeneous and stable. The world was thus depicted as a vast archipelago of cultures, each possessing its own internal logic and its own values, and which could exclusively be understood in its own unique terms. Variations in morality, custom and tradition were thus regarded as evidence of people's ability to adapt to the most variable environments and to shape their existence in a multitude of ways, and it was emphasised that there was no 'objective' standard available for the evolutionary ranking of cultures or the moral evaluation of actions. Value was defined from within. This fine of thought, which is historically associated with the great German-American anthropologist Franz Boas, is usually spoken of as cultural relativism or historical particularism.

Recently the classic perspectives from cultural relativism have become increasingly problematic, and cultural theory from the 1980s and 1990s tends to emphasise (now approaching the point of irritating reiteration) that cultures' are neither clearly bounded, tightly integrated nor unchanging. An important contributing cause, or at least a major catalyst, in bringing this change about, is the intensification of the globalisation of culture since the Second World War. The globalisation of capitalism and the modem state, along with innovations in communication technology, have been crucial for these changes to come about. When former 'tribals' now apply for mortgages, follow North American TV series, take their Higher School Certificates, elect local governments and are imprisoned for criticising the government, it becomes intellectually and morally indefensible to seek refuge in the fiction of assuming that cultures are isolated and committed to their 'proper logic': political discourse has, to a great extent, become globalised.

The situation may be even more problematic to handle intellectually for persons steeped in Boasian relativism when very tangible expressions of global cultural variation suddenly appear at our doorstep, which indeed is happening in most industrialised societies due to labour migration and to the ongoing influx of political refugees. This new polyethnic situation has, especially in European countries, provoked discrimination as well as a revitalised cultural nationalism and chauvinism in segments of the majority, but many -'indigenes' as well as new arrivals - have also responded by developing ideological and practical models for polyethnic coexistence. Original alloys mixing anthropological cultural relativism, nationalism, modem individualism and human rights thought have thus, in the course of the past 20 years, created ideologies and theories dealing with 'multicultural society'. In this milieu of social and political thought, difference is seen not only as politically legitimate, but is also frequently invoked as justification for specific political rights. In this regard, multiculturalist thought could be seen as post-nationalist, since it acknowledges the existence of several 'cultures' within one and the same political system. At the same time, multiculturalism may easily conflict with values seen as universal in modem liberal states, especially those to do with human rights and the rights and duties associated with equal participation in the institutions of society.

The basic dilemma of polyethnic societies can be phrased like this: on the one hand all members of a liberal democracy are (in principle if not in practice) entitled to the same rights and opportunities. On the other hand, they also have the right to be different - and in the 1990s, the rights of minorities to maintain and promote their cultural specificity, and to be visible in the public sphere, including the media, school curricula and so on, are increasingly insisted on. A crucial challenge for multi-ethnic societies therefore consists in allowing cultural differences without violating common, societally defined rights; in other words, the challenge consists in finding a viable compromise, for the state as well as for the citizens (representing power and agency, respectively, in the framework proposed in Wilson's Introduction), between equal rights and the right to be different.

This contradiction is as old as nationalism itself. Nationalism, the ideology holding that states ought to be culturally homogeneous, has a double origin in German Romanticism and French Enlightenment thought, which emphasise, respectively, cultural (in many cases ethnic as well) uniformity, and shared territory and citizenship, as the basis for national integration and as the source of political legitimation. According to classic Enlightenment thought, there existed a universal human civilisation, which was in principle accessible to all humans. According to German Romanticism, represented in the works of Herder above all, every people (Volk) had its proper linguistic and cultural character and the right to defend it. This view of culture, incidentally, was developed largely as a defensive response to French universalism, which was locally perceived as a form of cultural imperialism (probably not without a certain justification). This perspective and its derivatives (including cultural relativism in its 'strong' variants) are currently expressed through ideologies arguing the importance of cultural homogeneity for political identity. This applies whether they are nationalist and champion the idea of homogeneous states, or ethnopolitical and insist on ethnically based rights for minorities within existing states. However, the difference between 'German' and 'French' nationalism, so often stressed in the literature, is not absolute: in actually existing nations, the two principles are generally mixed, and even in principle, French territorialism is far from being culturally innocent. Insofar as the French universalist civilisation insists on speaking French, it has certainly not been perceived as culturally neutral among non-French speakers in Brittany, in Cote d'Ivoire and elsewhere. Modem human rights thinking is no more neutral either, incidentally, as it assumes global sharing of a specified set of societal values.

The contradiction between the demands for equal rights and for the right to be different is accentuated at present by two main tendencies. First, it has finally become dear in public discourse—nearly 80 years after Woodrow Wilson famously announced the right to self-determination of peoples—that hardly any ethnic ~;roup has its territory by itself. States are polyethnic, and any ideology stating that only people 'of the same kind' should live in a country is potentially dangerous. This problem was recognised already by Renan, but it has acquired unprecedented importance since the 1960s. Second, the current processes of cultural globalisation break down cultural boundaries and make it difficult to defend the idea that a 'people' is culturally homogeneous and unique. Cultural creolisation, migration and increased transnational communication are important keywords here.

A widespread counter-reaction against the perceived threat of boundary dissolution through globalisation consists in ideological emphases on 'cultural uniqueness'. In this sense, cultural homogenisation and ethnic fragmentation take place simultaneously; they are consequences of each other and feed on each other in dynamic interplay.

In other words, modem societies are by default 'multicultural' -or so it may seem. I shall nevertheless argue that 'multiculturalist politics' have to be universalistic in their very nature. The position to be defended below argues that culture is not a legitimating basis for political claims, and that cultural singularities among minorities and majorities in modem societies can only be defended to the extent that they do not interfere with individual human rights. All societies are indeed 'multicultural', whether they contain diverse ethnic groups or not, since different citizens hold different values and different world views. Multiculturalism, a term describing doctrines which argue the importance and equivalence of cultural heritages and the decentralisation of defining power as to what is to count as one, may in practice be a strong form of individualistic thinking about personhood - the world seen as a smorgasbord of identity options, This fine of thinking, described as 'critical multiculturalism' by Turner, may enable diverse groups to participate in public life, with their 'cultural heritage' or 'cultural identity' as their legitimated symbolic capital, and it can thereby be liberating for formerly silenced or unarticulate groups, including non-ethnic groups such as women and homosexuals. If, on the other hand, institutionalised differences form the core of multiculturalist practices, it is liable to regress into nihilism, apartheid and/or the enforced ascription of cultural identities. As the empirical discussion below will argue, the former alternative has many virtues in relation to human rights, which the latter does not.

Culture, Raymond Williams has written in a much quoted passage, is one of the two or three most complex words of the English language. The meaning of the word, Williams shows, has gone through many changes since the original Latin colere, which referred to the cultivation of the soil. Today, the word has several, if related, meanings.

One of the most common meanings of culture posits it as synonymous with the way of life and world view the members of a particular group or community have in common, and which distinguishes them from other groups. This definition may at first seem plausible, but it does not survive closer scrutiny. Within nearly every 'group' or 'people' there are varying ways of life and world views; the rich differ from the poor, the men from the women, the highly educated from the illiterate, the urban from the rural and so on. Additionally, it is often extremely difficult to draw boundaries between 'cultures'. If one argues that a Norwegian culture exists and is by default different from Danish culture, one will need to show what it is that all Norwegians share with each other but not with a single Dane. That is not an easy thing to do. Finally, culture is naturally not a solid object, even if the word unhappily is a noun. Culture is something which happens, not something that merely exists; it unfolds through social process and therefore also inherently changes.

Problems of this kind have made such a conceptualisation of culture difficult to manage, and many scholars have ceased to use it, while others insist on using culture in the singular sense, seen as that which all humans have in common, defining them as a species as opposed to nature in general and other species in particular.[2]

However, ideologists and political entrepreneurs of many shades have embraced this Romantic concept of culture. In recent years, ,culture' and 'cultural identity' have become important tools for the achievement of political legitimacy and influence in many otherwise very different societies - from Bolivia to Siberia. It is used by political leaders of hegemonic majorities as well as by the spokesmen of weak minorities.

Indigenous peoples all over the world demand territorial rights from the states in which they live, emphasising their unique cultural heritage and way of life as a crucial element in their plea. Immigrant leaders in Europe occasionally present themselves as the representatives of cultural minorities, demanding, inter alia, special linguistic and religious rights. The hegemonic elites of many countries also refer to their 'national culture' in justification of warfare or oppression of ethnic minorities. 'Cultural pleas' are, in other words, put to very different political uses.

A frequently mentioned 'paradox' concerning the break-up of Yugoslavia and subsequent war is the fact that the fighting parties, Serbs, Croats and Bosnian Muslims, are culturally very similar, yet justify their mutual hatred by claiming that they are actually profoundly different. This kind of situation, where ethnic relations between groups which are culturally close take on a bitter and antagonistic character, is more common than is widely assumed. In Trinidad, in the southern Caribbean, the following development has taken place in recent years. The two largest ethnic groups, Africans and Indians (originally from India; they are not American Indians), have gradually acquired more and more in common, culturally speaking; in terms of language, way of life, ambitions and general outlook.

At the same time, they have become ever more concerned to express how utterly different they are; culture and cultural differences are spoken about more often, and cultural differences are brought to bear on daily life, public rituals and political organisation to a greater extent than was earlier the case. Partly, this is because the groups are in closer contact than earlier and compete for the same scarce resources; but it is also partly because members of the two groups feel that their cultural boundaries are threatened by tendencies towards creolisation and therefore feel an acute need to advertise their cultural differences.

The groups have simultaneously become more similar and more different. This paradox is characteristic of globalisation processes, whereby differences between peoples are made comparable and therefore come to resemble each other, and where 'small' differences are 'enlarged'. It could, in line with this, be said that the entire discourse over 'multiculturalism' is embedded in a shared cultural framework encompassing, and bringing out the contradictions between, the Romantic notion of culture and the Enlightenment notion of individual rights. To put it somewhat more crudely: to make demands on behalf of a self-professed 'culture' indicates that one subscribes to a shared global political culture. The logic of multiculturalism and ethnopolitics shares its dual origins with the logic of nationalism in the Enlightenment and Romantic thought of early modem Europe.

In order to illustrate and further develop the preceding points, I shall now turn to an extended empirical example, which brings out many of the tensions and contradictions inherent in ideas of multiculturalism.

Since Mauritius was permanently settled by French planters and their African and Malagasy slaves in 1715, this island in the southwestern Indian Ocean has been a polyethnic society, and still is very much so, as witnessed in official symbolism as well as many aspects of everyday life. The currency is the rupee, and the text on the bank notes is in English, Hindi and Tamil. However, Mauritian newspapers tend to be in French, but the video shops offer mostly Indian and East Asian films. A leisurely walk through the capital, Port-Louis, may bring one past, within half an hour or so, a Buddhist pagoda, a Sunni mosque, an Anglican church and a Catholic one, and two Hindu temples - one North Indian, one Tamil. And it is by no means uncommon that Mauritians have names like Francoise Yaw Tang Mootoosamy.

Contemporary Mauritius, with a surface of some 2,000 square kilometres, has about a million inhabitants. Their ancestors came from four continents, and they belong to four different 'world religions'. According to official categories, the largest ethnic groups are Hindus from North India ('Hindi-speaking, 42%), 'Creoles' of largely African descent (27%), Muslims of Indian origin (16%), Tamils and Telugus of South Indian descent (9%), Chinese (3%), gens de couleur (2%) and Mauritians of French descent (2%). Mauritius, independent since 1968 and a republic since 1992, is a liberal multi-party democracy and a capitalist society (meaning, in this context, that both labour and consumption are mediated by money) which was impoverished, relatively overpopulated and dilapidated, with a vulnerable single export economy (sugar cane) and a high level of unemployment during the first decades after the Second World War. Mauritius has undergone an astonishing economic transformation since the early 1980s, and is now a relatively prosperous society with a dynamic economy based on sugar, textiles and tourism.

Mauritius is one among many peaceful polyethnic societies in the world. Although many of the country's inhabitants are concerned with their cultural identity, their 'roots' and the maintenance of local ethnic foundaries, compromise and tolerance are important ingredients in the shared Mauritian political culture. Notions which form part of a shared cultural repertoire include the admission that it would have been impossible to win a civil war, that secessionism would have been absurd, and that the country's political stability rests on a precarious balance between ethnic group interests. Therefore Mauritians have developed many more or less formalised methods for the maintenance of this balance.

Ever since France lost Mauritius (then Ile-de-France) to Great Britain during the Napoleonic wars, the recognition of difference has been an explicit tendency in its politics; first vis-a-vis the French settlers, since the 'niggers and coolies' were not initially endowed with rights. When the French capitulated in 1814, the Britons guaranteed the settlers that they would be allowed to retain their religion, their language, their customs and their civil rights. That the British kept their promise is evident today, as Mauritius is still much more French influenced than Anglicised. Even the legislative system appears as a unique blend of British law and the Code Napoleon.

During the twentieth century, and particularly since the extension of the franchise after the Second World War and the accession to full independence in 1968, policies relating to inter-ethnic tolerance have been extended so as to include the entire population. There is a continuous search for common denominators in legislation and in everyday social life, which are necessary for societal and national integration to be at all possible ('multicultural' or not, people need to have something in common if they are to have a society), and those universalist principles are balanced against the alleged conventions and culturally specific rights claimed by certain members of each constituent group.

The electoral system in Mauritius is more or less a carbon copy of the British Westminster system, with simple majorities rather than proportional representation.

The parties are largely organised along ethnic lines, and very many Mauritians vote for politicians who they feel represent their ethnic (sectional) interests. Attempts at creating Inter-ethnic alliances or supra-ethnic alternatives (based on, for example, class) have generally been short-lived. Although ethnic competition is in this way thematised in politics, there is nevertheless wide agreement over the political rules, and electoral results are respected. The Creoles, who are Christians, and the Muslims accept being governed by Hindus, who are politically dominant by virtue of numbers. At this level, there is no 'multiculturalism'. There is a shared discourse through which cultural variation and political disagreements may be articulated.

An important element in the Mauritian political system is the so-called Best Loser arrangement, which guarantees the representation of all ethnic groups through allotting a limited number of parliamentary seats to runners-up at General Elections. The 'best losers' are selected so as to ensure the representation of all ethnic groups in the Legislative Assembly. In this way, the importance of ethnic differences is made an integral part of the electoral system. As in many other multi-ethnic societies, questions concerning schooling, religion and language are among the most complicated and controversial ones in Mauritius. It is perhaps here that the dilemma of equal rights and cultural differences is most evident. In all three fields, compromises of various kinds have been developed.

Regarding religion, the popular idiom Sakenn pé prié dan so fason ('Everyone prays in his/her own way') has nearly achieved legal status. As mentioned, four 'world religions' are represented in the island, and three of them (Christianity, Islam and Hinduism) are divided into a large number of sects and congregations. Religious groups receive state funding according to the size of their membership. In this field, a consistent compromise has been established, where no religion is given priority by the state.

The Mauritian schooling system represents a different kind of compromise. Here, equality is emphasised rather than differences. Thus core curricula are uniform island-wide, as are exams. However, classes in 'ancestral languages' are offered as optional subjects. As a matter of fact, a growing majority of Mauritians speak Kreol, a French-lexicon Creole, as their first language, and scarcely know the language of their ancestors, but Kreol is rarely written. It could be said, therefore, that Mauritian schooling stresses equal opportunities yet allows for the expression of symbolic differences. It represents a compromise not only between ethnic groups, but also between a Romantic and an Enlightenment view of society.

A third kind of compromise is expressed in language policies. Officially, as many as fifteen languages are spoken in Mauritius; in practice, at least four or five are the mother-tongues of various groups. When Mauritius was to become independent from Britain in the late 1960s, in practice the new government faced four possibilities. First, it could have opted for Hindi, which is the ancestral language of the largest ethnic group (although many Mauritian Hindus do not understand it). Second, it could have chosen Kreol, which, in spite of its being held in low esteem, is by far the most widely spoken language. Third, French could have been an alternative, having been the dominant written language throughout the history of Mauritius.

In the end it was the fourth alternative, English, which was to win. English is an international language, and is learnt by Mauritians in the same way as non-native speakers elsewhere in the world learn English as a foreign language. This means that most Mauritians master it only partially. More importantly, perhaps, English was nobody's ethnic language, the few Anglo-Mauritians (most of them colonial civil servants) having either returned or become assimilated into the Franco-Mauritian group. By choosing English, an ethnically neutral language, as the language of the state, Mauritians avoided turning nation-building into a particularistic ethnic project at the beginning.

The other languages are nevertheless also supported through the state and its agencies. Public radio and TV broadcasting alternates between the major languages of Mauritius, and French still dominates in the written mass media. North American films are dubbed in French. There is in other words a clear, but negotiable division of labour between the non-ethnic language English, the supra-ethnic languages Kreol and, to some extent, French, and the ethnic languages, chiefly Bhojpuri/ Hindi, Urdu, Tamil, Mandarin and Telegu.

This will have to do as a general introduction to public policies relating to ethnic differences and national cohesion in Mauritius. The Catholic priest and ecumenic Henri Souchon became famous domestically when, at the height of the legendary 'race riots' of 1968, he admonished his congregation to visit the nearby mosque in order to familiarise themselves with a Muslim way of thought and thereby mitigate the mutual suspicion between Christians and Muslims.

More than two decades after the riots, Souchon sees two possible scenarios for Mauritius regarding the relationship between ethnic boundaries and the formation of identity categories oblivious to ethnicity.[3] He calls them the fruit salad and the fruit compote, respectively. In the fruit salad, the components are clearly distinct; ethnic boundaries are intact, and reflexively 'rooted' identities are secure and stable. In the fruit compote, on the other hand, the different fruits are squashed and mixed together with substantial use of force. (This metaphor, it may be noted, is a variant of the American melting pot metaphor.) The result of the compote de fruit, in pare Souchon's view, would be uprootedness, nihilism and confusion. He himself therefore supports the fruit salad variety, although he goes further than most in expanding the compass of the common denominators or, to stretch the fruit salad metaphor a bit, thickening the syrup. In order to have a dialogue, Souchon argues, one needs a firm position to conduct it from.

Some kind of fruit salad metaphor, or a rainbow metaphor which politicians are fond of invoking, is hegemonic in Mauritius. Yet conflicts between equality and difference are inevitable since the tension between sharing and difference is endemic to the island. Allow me to outline a few examples. Most Mauritian schools are public, but private schools also exist, many of them run by religious organisations. There are anti discrimination laws. It is nevertheless well known that Catholic schools have tended to prefer Catholic applicants for teaching positions, although they have also occasionally hired Muslims and Hindus. This policy was tried in court when an unsuccessful applicant filed a suit against a Catholic school in 1989 because she suspected having been by-passed on religious grounds.

In court the following year, the defence argued that it was necessary to have faithful Catholics in certain teaching jobs because a part of their job consisted in turning the pupils into good Catholics. The prosecutor asked whether this was also relevant with respect to subjects such as French, English and mathematics, which the school's lawyer admitted was not the case. In his testimony, the Archbishop, Mgr Jean Margéot, argued that the colours of the Mauritian rainbow had to be kept separate 'for the arc-en-ciel to remain beautiful'. The Catholic school won the case, and succeeded in this way in creating a precedent for differential treatment on religious grounds in a limited part of the labour market. The principle of difference here won over the principle of equality. Another nationally famous case from the same period concerned the controversial Muslim Personal Law (MPL), introduced during British rule, which allowed Muslims to follow customary Muslim law in family matters. A characteristic consequence of this law was that it became nearly impossible for women, but relatively easy for men, to obtain a divorce. In the course of the investigations of a Commission of Inquiry set up in the mid-1980s, it became clear that the opposition to the MPL was significant even among Mauritius's Muslims. Not unexpectedly, many women and young Muslims were against it, arguing that they were entitled to the same rights as other Mauritian citizens. In the end, the law was abolished, and universalist (Enlightenment) principles won over multiculturalist (Romantic) ones.

This second example is the most interesting one in this context. Here, the fundamental paradox of multiculturalist ideology becomes highly visible: it presupposes that the 'cultures' are homogeneous and 'have values and interests'. The mere fact that the formal leaders of an ethnic group invoke particular values and traditions does not imply that all members of the group support them. This is why it can be dangerous to accord special rights to groups, for groups inevitably consist of persons with often highly discrepant values and interests.

A third example highlights the relationship between particularist identities and universalist principles in a somewhat different way. Some intellectual Mauritians, tending towards the 'fruit compote' as an ideal, have experimented with mixing religions and cultural conventions in novel way, such as the radical music group Grup Latanier, which performs an essentially Creole séga music with strong Indian elements.

One leading Mauritian intellectual decided, some time during the 1980s, to challenge the rigid boundaries between different religions, reasoning that the island needed a 'shared culture' for a proper national identity to come about. On Christmas Day, therefore, he went solemnly to church, bringing bananas and incense as a sacrifice to the Hindu gods. This act was, naturally, frowned upon by Hindus as well as Christians, who both felt insulted by the blasphemous syncretism implied. If anything, they felt further apart after the experiment than before it. The ideal of the 'fruit compote' thus cannot be enforced against people's wishes. It should nevertheless be noted that universalist principles have been adopted by the Mauritian population with respect to political culture. In so far as discrepant religious or otherwise cultural practices do not interfere with the universalism guaranteeing individuals equal rights, there is no good reason to chastise them.

The Mauritian attempt at creating a synthesis between liberal principles of individual equality and a cultural relativist principle is remarkable and unusual, and it certainly deserves international attention.

The examples sketched in the previous section suggest that both equal rights and the right to be different may in particular situations lead to discrimination and the violation of commonly agreed upon individual human rights. If one insists on shared civil rights as the basis of citizenship and nationality, as the French revolutionaries did, one will tend to oppress minorities by forcing them to assimilate to a public culture (language, rules, hierarchies and conventions) which they perceive as alien and intrusive. If, on the other hand, one opts for differential treatment on the basis of religion or ethnicity, the risk is the opposite: those affected may lose their equal rights. South African apartheid policies are a good example of this; South Africans were encouraged to use their vernacular languages at all levels, and the majority of blacks were thereby in practice excluded from national and international political discourse. The hidden variable in this puzzle is, naturally, power discrepancies.

Additionally, it should be pointed out that political leaders and others are frequently prone to exploiting notions about cultural uniqueness strategically to strengthen their positions. In a critical study of ethnopolitics in the US, Steinberg concludes that persons and organisations generally invoke principles of cultural relativism when they themselves have something to gain from differential treatment, and that they will otherwise support equality principles. 'Tradition', 'rooted culture' and similar catchwords are positively evaluated in the political discourse of our time, and are often used rhetorically to justify privileges and political positions. On the other hand, this warning should not be taken to mean that there are never legitimate reasons for wishing to protect oneself against cultural domination. We just need to distinguish carefully between the right to a cultural heritage and particularistic politics.

Another, related point, which is also relevant for all polyethnic societies, concerns identification with collectivities in general. As a matter of fact, many Mauritians generally feel quite at ease as members of what they see as an emerging' fruit compote', and do not long for roots and purity. They would prefer to be cultural hybrids to the extent they wish, to be recognised as individuals and not as the representatives of a particular group. The legitimacy of this kind of strategy was tried out by members of the small radical socialist party Lalit ('The Struggle') before the General Elections of 1991. The militants on the list first refused to register their ethnic identity (which is compulsory, partly because of the Best Loser system), arguing it was irrelevant, and then proceeded to draw lots deciding their ethnic identity. The result was not devoid of Theatre of the Absurd qualities. For example, one of their leaders, to all appearances a white Mauritian of foreign birth, turned out to be a Hindu on the election rolls.

The neo-Romantic ideological climate influencing many parts of the world today - either viciously nationalist or equally viciously multiculturalist - is such that persons may virtually be forced to take on an ethnic identity whether they want to or not. Indeed, authoritarian culturalism may be just as oppressive in an ostensibly multi-ethnic and tolerant 'rainbow society' as in an ethnically hegemonic nation. The right to have an ethnic identity must also include the right not to have one. Here, perhaps, lies the greatest paradox of multiculturalism: in its apparently benevolent focus on 'the wealth of cultures and traditions' present in society, it neglects the Salman Rushdies of the world, so to speak; those persons who spend their entire lives midway between Bombay and London without wishing to, or indeed being able to, land. It excludes the 'mongrels', anomalies and idiosyncratic individuals who are numerous and necessary as interethnic brokers and in the forging of cross-cutting or non-ethnic alignments, and who represent the possible future of many societies.

Finally, cultural relativism gives no moral advice. To make it the source of public morality would imply that any practice would be acceptable as long as it can be justified by reference to 'a culture'. This kind of position is tantamount to no position at all, and is an inherent danger of falling into communitarianism.

It has often been asked why Mauritius is such a stable democracy, incorporating, as it does, a vast number of religious groupings and people originating from different continents. The question is wrongly asked, and it reveals an inadequate understanding of culture. At the level of everyday representations and practices, Mauritian culture can actually be described as quite uniform in the sense that there is a wide field of shared premises for communication encompassing most of the population: there is a shared political culture and a standardised and standardising educational system, there is considerable linguistic uniformity, and recruitment to the labour market is increasingly based on individual skills. It is generally not difficult to argue the virtues of individual human rights among Mauritians; they tend to share similar, Western-derived notions of justice. It is, in other words, only superficially (if noisily) multicultural even if it may be profoundly multi-ethnic.

It should be noted that the 'multiculturalist' model of coexistence, as practised in Mauritius and elsewhere, collapses unless the constituent groups share basic values of individualism and, in all likelihood, a shared lingua franca. For instance, it is widely believed, not least in that country itself, that the US has been capable of absorbing a great number of different nationalities without homogenising them culturally. This is wrong, and generally, migrants to the US have changed their language within two generations. One could perhaps say that the descendants of late nineteenth/early twentieth-century immigrants to the US have been assimilated to a degree of 99%, and have been allowed to use the remaining 1% to advertise their cultural uniqueness, which exists largely as a set of symbolic identity markers. As a Norwegian from Norway, I have often met Americans who identify themselves as 'Norwegians' but who unfailingly seem to betray, in their verbal and nonverbal language, lifestyle and values, a strong attachment to the moral discourses of US society.

If political multiculturalists favour equal individual rights, the /culture' in their rhetoric is but a thin cosmetic film. If, on the other hand, they seriously defend the right of ethnic minorities to run their own political affairs according to a cultural logic of their own, they run the risk of defending practices which conflict with the human rights of individual group members.

The solution, or rather, the 'good multiculturalism', must arrive at a blend of sharing and difference. It requires common denominators in key sectors, including politics, education and the labour market, and it must institutionalise a dialogic principl enabling a variety of voices to be heard on an equal footing. This is not relativism, but rather the recognition and democratisation of different value orientations in society, in the manner acknowledged as necessary and non-relativistic by Bauman when he notes the ill effects of the attempts at extending the Western 'ethical code over populations which abide by different codes ... in the name of one all-human ethics bound to evict and supplant all local distortions'. It is a question of striking a proper balance between the demands for formal equality and the demands for justice in a more general sense, including the equivalence of cultural heritages as well as the right not to acknowledge a heritage. The keyword is dialogue, which, it should again be noted, presupposes the existence of common denominators or shared meaning at the outset.

In the foregoing discussion, we have argued the importance of universalist human rights in modem state settings, and have alleged that political multiculturalism is a very fuzzy concept as it presupposes, yet explicitly and self-contradictorily resists, the presence of powerful processes of cultural integration. The very statement 'I have a culture worthy of protection' betrays a considerable degree of integration into a modem, reflexive way of thinking about the individual, human rights and politics.

Debates about indigenous notions of personhood in anthropology have frequently oscillated between positions stating, on the one hand, that remote peoples are 'just like ourselves'; and, on the other hand, that they are qualitatively and fundamentally different. Of course, both positions can be defended convincingly, given the appropriate selection and interpretation of empirical material. Regarding human rights issues, it is an often debated question whether or not they are or ought to be universal, and if so, whether they should be 'adapted' to local circumstances because of socioculturally conditioned differences in the constitution of the person. Be this as it may, the situation in societies where there are still groups which have been spared the mixed blessings of individualism, is not similar or directly comparable to the situation in Mauritius, the US or other thoroughly modern 'multicultural' societies where personal autonomy is considered an absolute value.

As many anthropologists have shown, concepts of personhood vary dramatically cross-culturally. In India and Melanesia, for example, a dominant view on the individual emphasises that he or she is a product of social relations and far from that self-sustaining, independent and inviolable 'monad' the Western individual is seen as. In such societies, the community rather than the individual is accorded rights, and the individual has duties rather than rights. In such societies, individual human rights can be seen as truly alien, even if they are often promoted and adopted by some segments of society, usually educated middle-class elites.

In his very beautiful and melancholy book Danubio, Claudio Magris writes that a fascist is a person who has best friends but cannot understand that others may be just as good friends; who feels love for his homestead but cannot understand that others may feel the same kind of love for theirs; and so on. It may therefore be proposed, as a general principle, that 'human rights missionaries' have an obligation to gain some understanding of the world views and value systems current among their target groups. They would then discover that virtually all peoples are, like Mauritian Muslims, divided on important issues. Some of their members would have gone to school and acquired individualist categories; some would have learnt about women's rights in remote countries; some might see a solution in a Marxist revolution or a liberal multi-party system, and yet others might refuse to question tradition.

As Samir Amin has written, individualist thinking and social criticism is just as 'rooted' in Islamic history as fundamentalism. And as Salman Rushdie and others have reminded us, one scarcely does southern or eastern peoples a favour by continuously telling them that individual human rights are really a 'Western' invention and far from an aspect of their culture. This kind of attitude essentialises 'other cultures' and alienates the growing numbers in those societies which hold positive views of individual human rights at the same time as they resist cultural neo-colonialism.

Integration in a modem state with a liberal constitution may create a dialogical situation where human rights principles become a common denominator for the many groups and individuals which make up the state. If this sounds like blunt cultural imperialism, it should be noted that the most likely alternative, in my view, consists in a form of segregation whereby the exertion of power is left to persons such as the old men who are the formal leaders of Mauritian Muslims, and where there is a mounting risk of ethnic conflict because of the inter group competition implied by segregation.

In most contemporary societies, processes of cultural homogenisation are taking place in some social fields (such as consumption, education and the media), while the demarcation of boundaries and the symbolic strengthening of 'identity', 'roots' and 'tradition' takes place in other fields. It is this process I described at the beginning of this article as the dual movement of cultural homogenisation and ethnic fragmentation. In this context, the Mauritius I have described may perhaps serve as a microcosm or an ideal type of a modem society: Mauritian society is simultaneously characterised by conflicts and contradictions, pluralism and value conflicts along several axes, and one cannot offhand say what kind of values or morality 'society as such' represents. For this kind of society to be cohesive at all, common denominators are necessary, and a recognition of cultural diversity which does not interfere with the principle of universal, individual human rights may actually be the best alloy available. It is the blatant non-recognition of cultural heritages which leads to ethnic revitalisation and fundamentalism, not their institutionalisation through the state.

India was mentioned above as an example of a society where Western human rights thinking seems outlandish and alien. It might therefore be appropriate to end by stating that Marxist, feminist, liberal and other kinds of individualist human rights related movements enjoy great support in Indian society. Such groups are no less 'authentic Indian' than those of traditionalists who dream of a reawakened Hindu millenarian kingdom where ancient hierarchies are respected in minute detail.

Perhaps it would be useful to speak of a 'weak' and

a 'strong' variant of political multiculturalism. The former is the one practised in some liberal modem states, including Mauritius, where a high degree of cultural homogeneity is taken for granted. The latter, which I have argued against, would be a kind of political rhetoric rejecting liberal individualism and human rights ideology on the basis of alleged tradition. The former, 'weak' variety is, however, also hard to defend as a political project seen within a human rights perspective. It may, as have argued with reference chiefly to Mauritius, (1) contribute to freezing ethnic distinctions and thereby heighten the risks of ethnic conflict, (2) remove the protection and entitlement of shared societal institutions from the members of minorities, (3) strengthen internal power discrepancies within the minorities, (4) direct public attention away from basic contradictions in society, notably economic ones, and (5) contribute to a general moral and political disqualification of minorities in society: since they are not accorded the same rights and duties as everybody else, there is no apparent reason why they should be treated as equals in other respects either. The conclusion is, thus, not that cultural variation in itself should be combated, but that politicised culture is incompatible with the individual rights which modern states are, or ought to be, based on. The slogan could be 'cultural nationalism, political cosmopolitanism', to borrow a turn of phrase from Gellner.

13
Towns and Cities

Many centuries ago the invention of the plow as well as the domestication of animals made it possible to produce food in greater quantities and more efficiently. The invention of the wheel and the sail cut transportation costs and led to increased trade. The use of metals made it possible to process raw materials much more effectively and to create a variety of useful machines. At the same time more land was being put under cultivation, the population was expanding, and the needs of the people for goods and services were constantly growing, thus increasing the need for towns. In modem times the service function has become of great importance. The urban areas are geared to feed, clothe, shelter, govern, educate, entertain, finance, and manage-in fact, to provide any service which people demand and can pay for. All this is not to say that urban functions are limited to the towns and cities; factories, service industries, and many other urban units are constantly moving to country places which, if the movement continues, will themselves become urban.

Of great importance in the early expansion of the city was the proximity of a considerable area of fertile land where grain and other crops could be grown in abundance to provide a surplus which the city might gather together and transport to distant places in exchange for, other goods. Cities like Paris, which was established at the center of the Paris Basin, drew much of their support in the early days from their immediate environs. It was not until the last century that, Paris came to depend for its food upon distant places. Many other cities owe their first urban development pattern to fertile agricultural land, forests, or mineral deposits in the vicinity. Another important factor in the development of cities has been accessibility to trade routes, many cities having grown up at crossroads, sea ports, and river ports.

The unfolding of the modern city has been dependent upon the growth of new sources of energy or power. So long as man had not developed much water power or steam power and was confined to the use of human power and animal power, the city was limited in its function, particularly from the standpoint of manufacturing; also, transportation was necessarily slow, capable of accommodating only relatively small burdens.

Towns

The term town has different meanings in different parts of the world. In New England, for example, it is a governmental unit or subdivision. In parts of Europe, as on the outskirts of London, "town" refers to large service centers which are, in fact, cities. Towns grow up in specific places to satisfy specific needs and the particular location is often a key to the function.

In general, towns are concerned with storing, processing, buying, and selling the surplus products of the surrounding region, the hinterland. The town also distributes the goods imported from other towns and regions. In addition, the town has other service functions, including schools, doctors, dentists, lawyers, and, in some of the larger towns, a weekly newspaper. Some towns, such as resorts, have specific functions; here the usual storage and processing functions may be of slight importance. In some of the older towns the site was chosen from the standpoint of defence. Even after the necessity for defence has passed, the town continues in the same site but with a different function. Site is critical for the initial location of a town or city but it is the situation or hinterland which affects the growth and character of the settlement. Favourite town sites are where streams can be easily forded or where it is feasible to have a ferry.

The form of a town, that is, the pattern of the streets and buildings, depends upon a number of factors including the site, the nature of the transportation routes, and the ideas that people have for developing their town. Where there is restriction to building, as in a narrow valley or on a steep coastline, the town takes on an elongated, linear shape.

The pattern of streets is likely to reflect the nature of the transport routes; thus in a crossroad location the town might have a rude, rectangular pattern; newer towns, developed after the construction of railroads, often are oriented to the railroad rather than to a natural feature.

One of the most common street patterns is the rectangular grid, developed especially in areas where the land survey system is oriented in a northsouth direction, as in many parts of the United States. It is not uncommon to find a street pattern in the core of a town oriented to some natural feature, a stream or shoreline, and the newer, outlying parts of the town oriented to the land survey system.

In New England towns an open square, the "commons," is reserved near the center and in parts of southern United States the "courthouse square" is a prominent feature of the town. The radial circular pattern developed in some villages persists in the core of the town while the fringes may have an entirely different pattern. Towns located on a relatively level plain tend to sprawl out in every direction along the main lines of communication and the outline is roughly star-shaped.

Cities

Cities are to be found in all parts of the earth that have even moderately dense population. Most of the large cities, however, are located in regions with both dense

Hinterland

Important to every city is the tributary area or hinterland, the area for which the city is a focus. This is both the area served by the city and the area upon which the city is dependent. The city gathers the products of the tributary area and prepares them for either local use or export to others regions.

The city also imports products from other regions and distributes them to the hinterland. The area of a city's hinterland depends on a number of factors, including the size and character of the city, the nature of the goods produced, and the efficiency of the transportation system. If one factor is taken separately, it is possible to apply the central place theory.

Central place theory, and that portion of economic geography which considers the relations between primary agricultural production and its marketing, started with the contributions of Johann Heinrich von Thunen. Von Thunen assumed a single market town in the center of uniformly fertile and arable plain, accessible equally from all directions. He stated that the prices of farm goods in the market and the cost of taking the goods to the market would determine rather regularly how near to, or far from, the market these goods would be produced: in other words, the extent of the town's hinterland.

For the purpose of studying systematically the ways in which goods production and land use vary by distance from the market, we may rephrase von Thunen's verbal model: "Because market price equals the cost of production plus the cost of transport plus the profit to producer, land use will vary systematically with distance of the producing unit from the market point." Symbolically this reads: M = P + T + G, where M is market price, P is production cost, T is transport cost, and G is gain, or profit.

The easiest situation to deal with in logic or laboratory is one in which all possible complicating factors have been eliminated. After understanding how a model works under conditions of restraint or limitation, we can progressively approach reality by relaxing the theoretical restraints. For the present, the basic restraints on the von Thunen model may be assumed as follows:

1. A single selling point, the market.
2. A single commodity produced on each production unit.
3. Equal level of fertility everywhere.
4. Equal land transport facilities everywhere.
5. No political regulation of the economic factors in the model.
6. A constant market price for a given commodity.
7. No upper limit on the amount of arable land.
8. One political unit.

Since most cities have multiple functions, it is not easy to define the hinterland quantitatively. Within the hinterland of large cities, there are smaller cities, towns, and villages which take on some urban functions common to the larger city.

Generally, these are retail and service functions needed rather frequently by people, for which they are not willing to travel often to a large city far away. The drawing of hinterland boundaries becomes complicated, but it has pattern if we think of a hierarchy of central places. Since the hinterlands of large cities include many smaller cities and their hinterlands, there is, in fact, a hierarchy of trade areas from the hamlet to the metropolis.

Functional classification of cities

The cities of the world are so different that they must be roughly classified before their characteristics can be properly discussed. Harris classified United States cities as follows, using census data for gainfully employed persons with a few examples of each:

1. *Retail*. More than 50 percent of employed persons engaged in retailing. Most small cities of less than 100, 000 population.
2. *Manufacturing*. More than 60 per cent employed in manufacturing.
3. *Wholesaling*. More than 20 per cent employed in wholesaling.
4. *Transportation*. More than 11 per cent engaged in transportation.

In addition to the functions of a city indicated in Harris's classification, there are other functions performed by the city, particularly secondary functions that have to do with the support of the city itself. Although a city may be primarily engaged in trade, wholesale and retail, for the benefit of the surrounding area, it still has to take care of the people living in the city, in many ways its most important function. Typically, there is differentiation and specialization in the city.

This system of classification is obviously based upon major function, since nearly all large cities have some element of most of the classifications mentioned. Thus, a manufacturing city also has important retail wholesale functions, as wel' as functions characteristic of transportation centers.

Very important qualities of a city are its site and its situation. The site is the actual location of the city, the ground upon which the city is constructed; the situation is its general location with respect to the surroundings. Thus, the site of New York City is Manhattan Island and adjacent parts of the mainland and Long Island, while the situation of New York City is in northeastern United States at the mouth of the Hudson River. The site usually has great influence on the form of a city; the situation of a city and the characteristics of its trade; area, or hitherland, often have much influence on its functions.

It is easy to recognize the manufacturing district of the city, especially when it is composed of the more obvious forms of manufacturing. Heavy manufacturing carried on in steel mills, large automobile or airplane factories, and machine shops is clearly indicated by the form of the structures and by the activity that is going on.

Other kinds of manufacturing, especially various types of light manufacturing, are not so easy to recognize since many of them are carried on in the upper stories of buildings which house retail stores on the street level.

The central business district of the city is clearly distinguishable by its banks, whole sale and retail stores, and hotels. The whole sale section is usually distinct from the -retail, at least in part, being represented by large warehouses and a congestion of heavy transportation, including railroads, trucks, and occasionally water-borne carriers.

An important feature of every city is the residential district, where the people live to carry on the work of the city residential districts are characteristically peripeheral, there are some that are well within heart of the city. Usually, residential districts are pushed out into the rural areas as the city grows, and their former area is occupied by retail, manufacturing, or wholesale establishments. Although the functional parts of the city (the manufacturing, retail, wholesale, and residential districts) can usually be recognized by their structures, confusion may arise in areas of rapid growth.

The central business districts (CBD) of most cities are growing, although not as rapidly as the cities themselves. In many cases some of the retail establishments are moving out to the margins of the city where land values are lower and more parking space is available.

Usually on one front the CBD is advancing-the zone of assimilation; and on another it is receding-the zone of discard. The advance is usually toward the better residential areas, while the retreat is in the area of skid roads and slums. In the first instance many houses built for residences are being taken over for commercial purposes; in the second, structures built for commercial purposes are taken over by missions and as low-rent housing, or they are abandoned. The focus of the CBD is the peak land value intersection(PLVI).

World cities

The great cities have a complex functional pattern and, therefore, a complete structure. In the first place, metropolises are usually centers of political power, not only for the immediate region but for the state the nation, and, in some cases, for inter national governments. Gathered here an many professional organizations, the trade unions, employers' federations, and head quarters for major industrial concerns Cities plan for, manage, and finance their hinterlands. These cities are centers of world trade Many of them are great ports and all of them are well situated as transportation centers. Airlines, railroads, and highways focus on them. Many of the world cities have well-known airports: Kennedy Airport (New York City) and Orly Field (Paris). The cities are centers for universities and publishing houses, and also headquarters for the great national television and radio networks. Not only do many of the people of the world live in the large cities but nearly all of the affluent people live there. This has led to the development of high-quality entertainment and also to many luxury industries and shops including large department stores which cater to every demand of the wealthy residents. And the luxurious establishments such as opera houses, theatres, concert halls, and expensive restaurants, are 'now available to a very large share of the city population.

The manufacturing function of most of the large cities tends to be highly specialized. Few of them are noted for heavy manufacturing. As the economies of the countries have become more advanced, the emphasis on manufacturing in the large cities has shifted from what formerly were basic industries to highly specialized ones requiring skill, design, and attention to fashion, catering to the particular needs of the cities' residents. In other words, most large cities manufacture primarily for their own residents and not especially for a world market.

It was inevitable that a large city should grow up at the lower end of the Hudson River; both site and situation are favourable. The sites on Manhattan Island and the adjacent shores provide for the development of docks and other harbour facilities. The adjacent water is deep enough in general for the largest ocean-going ships.

The location at the lower end of the Hudson River gives New York easy access to the Great Lakes and to the interior of the continent, a quality which no other city on the eastern seaboard enjoys to such an extent.

The most striking quality of New York City is its great size. In many ways it is factory jobs. The need for planning, financing, advertising, selling, and promoting has led to the growth of a large, elite group of people in contrast to the factory workers. More efficient methods in factories have led to technological unemployment in some cases, but this is more than offset in the cities by the increase in white collar jobs.

The growth of the cities has brought with it a number of serious problems. The first is the problem of housing. When thousands of new people arrive in a city they must find houses, and in many cases slums have developed because of the congestion in the inner districts on the margin of the CBD. The demand for housing has resulted also in suburban sprawl, where adequate facilities of sewage, water supply, and recreation are not available.

New York City

The most striking quality of New York City is its great size. In many ways it is the largest in the world, no matter how the population is counted. If is defined as the New York—northeastern New Jersey standard consolidate area, it covers an area of nearly 1960 was 14, 700, 000. If the city id defined as the Tri-State New York metropolitan region it includes an even large area, 22 countries, one of which is in the state of Connecticut.

Manufacturing became important in Manhattan early and has continued ,to the present day. Although many types of factories and even the types of wholesaling that require storage of bulky commodities have moved away from Manhattan to the outer ring on the Jersey shore or on Long Island, some types of manufacturing persist, particularly ladies' garments and book publishing. In the latter case, the actual printing of the books tends to be farmed out not only to outlying districts but to other cities.

The women's garment industry has remained in Manhattan largely because it is subject to rapidly fluctuating-styles. It is important to every ladies' garment manufacturer to know what his rivals are doing so that he can copy their styles. In addition to the garment and book publishing industries, many other types of business activity, such as food processing, profit by a location in New York. It would be to their disadvantage to be located farther away. The number of office workers employed in Manhattan is constantly increasing, but at a slower rate, perhaps because of the widespread automation of office work.

As more and more space is used for business, less is left for residence. Nonetheless there are more than a million-and-a-half people living in Manhattan. The character of the population has changed too. The most recent changes involve a large immigration of Negroes and Puerto Ricans.

New York City is now the largest "Negro city" in the world and also the largest "Puerto Rican city." The Negroes and Puerto Ricans live in the most congested and poor residential districts in Manhattan. In 1960 there were 1,141,000 Negroes in New York City and 613,000 Puerto Ricans. This represents enormous increases in the case of the Puerto Ricans, 149 per cent over 1950. Manhattan still has large numbers of other immigrants. In the lower midtown section, especially on DeLancy Street, there is a concentration of Jews; west of this is the Italian quarter.

Formerly New York City was the largest "Italian city," but the Italian immigrants, as well as those from a number of other European countries, have in many cases moved to outlying districts. To the southeast is the Chinese quarter and to the northwest, the Polish and Ukrainian quarters. The tendency of a group of immigrants to live in the same neighbourhood has led to the statement that it is possible to "go around the world in New York." In an attempt to ease the conditions of slum living, especially for Negroes and Puerto Ricans, but also other low income groups, large-scale urban renewal projects have been initiated. However, the number of new housing units constructed is a very small proportion of those needed and, furthermore, the rental cost of the new units, low though it is often too high for the people who were displaced.

Most of the people of Greater New York live in what are ordinarily described as suburbs; in New Jersey, Connecticut, Long Island, and various other places. As in the case of many other American cities, the growth of the suburbs does not represent a continuous extension of the city, but rather the development of spots or areas at some distance from the city between which there are large areas of unused land, at least unused for residential purposes. This makes the problem of transportation even more serious. The transportation problem in New York City is serious but it could, perhaps, be worse. Many of the people who work in Manhattan live two or three hours away by public transit. Some take railroad commuter trains to the outskirts of Manhattan and then transfer to the subway to reach the vicinity of their offices or working places. On an average working day 1,600,000 people enter Manhattan in the area south of 61st Street, that is, south of Central Park; 225,000 of these have come by private automobile, taxi, or truck. The remainder use public transportation, buses, trains, and the subway to get to work. So, the use of automobiles in New York is much less, in proportion, than in Los Angeles or Chicago. Nevertheless, in downtown Manhattan, and in other areas in New York City, automobile traffic congestion is very serious, especially at the rush hours.

New York City not only has acute problems in the field of housing and transportation but also in pollution. The automobiles used in the outlying areas and, to a certain extent, in Manhattan create a hazardous atmospheric condition which is compounded by some of the refineries and factories in the outlying districts.

London

London is not only a great city, but it is also a great port and a great capital. In Roman times the small town, Londinium, was located at the lowest port on the River Thames where the river could be forded easily. Above this point the Thames is a small river, essentially unnavigable; below, it widens into London River, which leads to a broad estuary giving easy access to the continent across the English Channel. The port is favoured by a double tide, which means that the high tide period is of several hours duration, enabling large ships to come into docks which otherwise could not be reached. London lies in a gently sloped plain called London Basin. To the south and northwest of the city are low hills, but most of the country is of little elevation so that circulation on land by rail and highway is not handicapped by terrain, and the city has developed as a series of fairly regular concentric rings. The core-about 10 square miles in area-is located to the north of the Thames.

This is "The City," the central business district with many government buildings and historic sites. The resident population is only about 270,000 and, as in other large cities, is declining; but a work force of a million-and-a-half people commutes to The City during the day. Until recently the construction of office and other buildings in the core area was restricted to about 10 stories so that the skyline of the city appeared very even. Recently this rule has been relaxed and many skyscrapers have been constructed, with many more under construction. Adjacent to the commercial core of London is the "inner ring," a zone of industry and residence about five miles wide with a population density of about 27,000 per square mile. The inner ring is divided into the east and the west with very contrasting uses.

The East End has traditionally been the area of poor people and industries, particularly the chemical industries, the gas plants, and various other light and medium types of industries; London has never had a large amount of heavy industry.

During World War II the East End was bombed heavily and much of it has been reconstructed. The West End has traditionally been the home of the wealthy Londoners, with fine old palatial residences in pleasant surroundings. Because of the rapid expansion of population many of the old residences of the West End have been converted into flats (apartments) in which people are crowded in order that they may live comparatively close to the core.

Beyond the inner ring are the suburbs, 15 miles or less from the central business district, referred to as the Interwar Suburbs, since most of the growth here occurred between 1918 and 1939. This zone has fairly good transportation connections to the core by way of the underground railways and surface electrified lines. The highways are less satisfactory, crowded with a density of automobile traffic far greater than the builders imagined forty years ago. The Interwar Suburbs are losing population slowly as the original residents grow old and their children move away. Beyond the Interwar Suburbs is the Green Belt, a zone of open country about five miles in width in which building is no permitted (in Great Britain permission t build on agricultural land must be obtained from the national government). There is, of course, great commercial pressure to open parts of the Green Belt to building.

The creation of the Green Belt forced the expanding population farther out. To Meet this situation eight carefully planned New Towns with a total population of more than one-half million were built, 30 to 40 miles from the center of London.

These towns (which in fact are small cities) are complete urban units with their own offices and factories but they also function as residences of people commuting to the heart of London. Unlike comparable' suburbs of New York City, the New Towns are separate communities with open country between the urban units. The development of this outer ring is restricted to a few localities.

The problem of circulation in the London area, which has not been as serious in the past as in the case of some other large cities, is nevertheless becoming more acute because of the increased use of the automobile.

As long as most of the commuters travel by train, bus, and underground, the conditions are not too serious, but every year more and more people are using automobiles for commuting and particularly for travel during the evening and on weekends. The congestion will continue to increase unless a large number of offices and government bureaus can be moved from the center of London to outlying cities, and this is not easy to accomplish.

Paris

The original location of Paris was on two small islands in the middle of the Seine River in the heart of the Paris Basin. This first location was selected for defence and because it made it possible to the Seine by two narrow channels instead of one broad one. Paris developed as a walled city on both banks of the river; the first wall encompassed only about 675 acres of land. As a result of the wall building and the defence requirements, the streets were laid out in a radial and circular pattern. This gives the appearance of a planned city but Paris certainly was not planned in terms of its modern requirements. (The plan of Washington, D.C. with its radial streets and numerous hubs or circles was copied from Paris.)

Paris is not an ocean port, but it does have waterways that are very important to its life. The Seine meanders generally toward the northwest to the English Channel and connecting canals reach southeastward to the Rhone and the Rhine. Much of Paris's freight is carried on small craft in the rivers and canals.

Paris is the administrative, commercial, and manufacturing center of France to a degree not found in other large capital cities. All the important railway lines radiate from Paris to outlying cities such as Bordeaux and Marseilles. Peripheral railroad lines are much less developed. The commercial core of Paris, La Ville de Paris, is much more difficult to define than those of New York or London. It is a mixed area with many more residences than are commonly found in the commercial core of large cities. One way of defining the commercial core is to include the 10 central districts, or arrondissements. In this area of 40 square miles nearly three million people live. This represents a much greater density than in the case of London. Here 90 per cent of the dwellings were built before 1914 and in the east end where the poorer people live, many of the dwellings have no bath or shower. The concentration of population in the commercial core makes for a very great congestion in spite of a good transportation system of subways, surface buses, and rail lines. The railroad stations are on the margin of the commercial core and there is inadequate provision for moving people from the stations to the various parts of the commercial core.

Many years ago the streets of Paris were rebuilt and widened so that there are many wide boulevards, which are seen by all thc tourists; but almost no provision is made for parking, and with the increased use of the automobile the congestion is compounded. Many of the people of Paris have moved out to the suburbs.

The principal suburb of Paris is in the Department of the Seine and administratively it is not completely under the jurisdiction of the city of Paris. This corresponds to the suburban ring of London. In 144 square miles of this suburban area live 2,850,000 people; it includes individual houses, apartment houses, and the largest factories of Paris.

The factories in the inner core are mostly smaller; however, most of the manufacturing of Paris is concentrated near the waterways and especially near the Seine, both to the north and to the south of the commercial core. Among the comparatively heavy industries are sugar refineries, oil-pressing mills, railroad equipment shops, cement plants, automobile factories, -ceramic factories, and power houses to supply the city with electricity. In the inner core, for the most part, are the smaller industries such as the clothing factories, factories for perfume, and for many other products. The industries of Paris are highly diversified and produce almost anything needed in the city. Paris now has a master plan for the future growth of the city. This involves improving the transport system, moving some industries and offices out of the central area, and removing, demolishing, and rebuilding some of the ancient buildings in the city.

Several possibilities are open for improvement of the transport system or, as the French call it, the transport infrastructure. The subway (Metro) can be extended beyond the old limits of the city and into the suburbs, the surface transportation by bus can be more carefully coordinated with the Metro terminals, and provision can be made to move people from the railroad stations into various parts of the commercial core. Paris is one of the greatest of the great capitals because it so completely dominates, the country in which it lies. It has the reputation for charm, for beautiful living, and attracts millions of tourists.

Moscow

There seems to be no logical reason for the location of a large city on the small Moscow River, a tributary of the Oka which in turn is a tributary of the Volga; but it was inevitable tan an important city so 400 miles from the Gulf of Finland at Leningrad, 750 miles from the Black Sea of Odessa, the greatest of all Russian grew. Moscow A more than six million people in the city proper and more than ten million in the regional center. This includes a number of sputnik satellite towns, 20 to 40 miles from the center, of Moscow. It was a historical accident, perhaps, that the military princes who conquered this area centuries ago selected this particular place as their capital; in the long run it turned out to be a good choice. In the early years the city grew very slowly: in 1860 it had only about 360,000 people. The capital was moved from Moscow to St. Petersburg (Leningrad) for a period but it came back at the beginning of the modern revolution in 1917, and since that time Moscow has grown very rapidly indeed. The industrial advantages of Moscow are many.

In the vicinity there are deposits of coal and peat and various earth materials such as clays for ceramic industries, but characteristically the industries of Moscow have imported much of their raw material from other parts of the U.S.S.R. or even from foreign countries. This includes raw materials for textiles, petroleum for power generation and general use, and steel from the Donbas region to the south. The Moscow district, called the Central Manufacturing District, is the most important in the U.S.S.R. for its variety of industries. Moscow has the best railroad connections of any city in the U.S.S.R. It is on the main cast-west line, the Trans Siberian, and railroads also radiate in all directions from Moscow.

Undoubtedly a very large share of the importance of Moscow lies in its function as a capital of a great country. The administrative heart of the country is located here and in addition Moscow serves as an international center for Communism; other Communist countries look to Moscow as their great administrative and cultural center.

Moscow has come to have meaning far beyond that of the usual capital of a large country; it stands for an idea. One commonly hears the expressions, "Moscow thinks," or "Moscow says." This particular idea-function of a capital is remindful of some of the great capitals of the past of Rome or Constantinople. The interior problems of Moscow, like those of many other large cities, revolve around housing and traffic. The rapid growth of the population and the limited amount of, private building have created a very acute housing situation.

Until recently the housing units in Moscow averaged one room to every four persons. This is being improved by a crash program of government construction of five-story apartment houses, available at moderate rentals. Most of these apartment houses are made of pm fabricated concrete units assembled on the site, and they still do not represent the optimum in housing for city dwellers. The traffic problem is less serious, perhaps because of the small number of automobiles. In the U.S.S.R. as a whole there is about one automobile to every 200 people and although the ratio is much higher in Moscow, still there are fewer cars in Moscow than in London or Paris.

Tokyo

The city of Tokyo, located at the he d of Tokyo Bay and in the midst of the most productive area of Japan, the Kuant'ao Plain, is one of the largest cities in the world. Many think it is the largest but, depending on definition, New York City is sometimes rated as a little larger. The site is a favorable one: the broad bay gives considerable access by water and although the depth, is not naturally suitable, for large ships, the harbour of Tokyo has been deepened to accommodate medium-sized freighters. The largest ships still stop at Yokohama down the bay, which is essentially a part of the Tokyo metropolitan area. It is difficult to define the limits of Tokyo with any degree of satisfaction. The inner 23 wards have a population of over 8 million. The metropolitan area has a population of more than 10 million and the Institute of Urban Research definition of the outer limits gives it a population of over 14 million. Since the year 1615, the growth of Tokyo has been rapid; by 1795 it had a population of 1,400,000, which made it by far the largest city in the world at that particular time. Kyoto was the official capital of Japan until 1868, when Edo was renamed Tokyo, meaning eastern capital." The growth has been mostly by in-migration in recent years, since the birth rate in Japan has fallen substantially in the last 25 or 30 years, and more especially since World War II.

In spite of many disasters, earthquakes, fires, tidal waves or tsunamis, and war damage inflicted in World War II, the growth of this city has scarcely been interrupted or delayed at any time. Aside from being the administrative and commercial center of Japan, with all that goes with it, Tokyo is a very important manufacturing city. There are two main foci of manufacturing, one in the Kawasaki area where the heavier industries are concentrated: refineries, metal working, cement industries, chemical industries of various kinds; and the other in the northern part of the city where a great variety of manufactured goods are turned out: autos, trucks, buses, radios, television sets, cameras, many types of light metal goods.

Japan has been able to expand its industries very rapidly following World War II because of the highly developed skill of its labourers and the low cost of labour, plus the fact that the markets in many parts of the world, including that of the United States, have been open to Japanese goods with only moderate tariffs applied to them.

Tokyo has shared significantly in this expansion. The growth of Tokyo has been so rapid that in addition to problems of housing and traffic the acute problem of waste disposal must be added. The housing problem is made worse, perhaps, by the restrictions on building because of the earthquake hazard.

Most Japanese families prefer single-family residences and all of these are one or two stories in height, which in such a crowded city is wasteful of space. Many apartment complexes are being built but buildings are restricted in height to 102 feet, so that very high rise apartments are not permitted.

New construction is limited by the scarcity of land, which is very expensive, and the small amount of credit available to the ordinary Japanese citizen. Most of the rapid increase in population over 200,000 persons per year-is in the zone from 10 to 15 miles from the center of the city; this in itself creates a traffic problem. In the center only about 10 per cent of the area is in streets, as compared to about 35 per cent in New York City.

Buenos Aires

In the Southern Hemisphere there are comparatively few large cities. In South America Caracas, Lima, Santiago, Rio de Janeiro, Slo Paulo, and Buenos Aires are all over two million in population. Buenos Aires was established in the year 1536 as the gateway to Andean America. To reach Peru from Buenos Aires required a long journey by river boat and overland, but the alternative was to sail around Cape Horn or to make the difficult portage across Panama.

Buenos Aires was located on comparatively dry ground in what is in many places a swampy coast, but it was close enough to deep water for the sailing ships of early days. The water depth near the shore is not sufficient for modern ships so that an artificial harbour has been created with docks and two dredged channels which lead from the harbour into deep water. Also a new harbour has been established 35 miles to the southeast near the planned city of La Plata.

14
World Trade

Trade may arise whenever one area has a surplus of goods or services that are wanted by the inhabitants of another area. The second area may pay for these by complementary goods and services. Trade may also occur when a nation exports nonsurplus items in order to obtain a trade item which is of paramount importance to it.

Surpluses are generated by differences in the physical environment and the human environment. If an area has good soils, moderate relief, and an equable climate, it probably will have an agricultural surplus. Geological events of the past determine the nature of an area's mineral surplus. Since the differences in the physical environment are less readily changed than those of the human environment, trade ultimately rests primarily on differences in, resources. Of course, the factors in the human environment still have a certain degree of importance.

Differences in technical abilities and the economic momentum of a "head start" in industry are forces making for a real differences in surpluses. World trade involves the exchange of surplus goods and services between countries. Portland, Oregon, is a port city which is important in world trade. Ships come partially empty up the long Columbia River estuary and dock along the banks of its tributary, the Willamette. Here at Portland they unload the cargoes, which for the most part are manufactured goods. Much bulky raw material (lumber, wheat) is loaded, as are such processed goods as aluminum ingots, plywood, light machinery, and canned fish.

Trade within a climatic zone depends less upon differences of physical environment side than upon geographical division of labour. This division is due to differences in population density, historical background, and level of technology. Since climatic zones tend to parallel or be subparallel to the Equator, we may say that trade moving within a climatic zone or production region tends to be based primarily on differences in human environment.

Trade moving between climatic zones tends to be based more firmly upon differences in physical environment. A good example of trade within a climatic zone is the rice trade in Asia. Heavily populated China, Japan, and India are large rice producers, but they consume all of their production. The more sparsely populated countries of southeastern Asia until very recently produced a surplus of rice, which moved outward to the more densely populated areas. Burma sold to India and Ceylon, Thailand to Java and Japan, and Vietnam to Malaysia and Java. Thus, within the area in which rice can be grown, surpluses in one place and need in other places gave rise to foreign trade.

Trade generated by differences in physical environment is illustrated by the ex change of commodities between tropical and mid-latitude areas. For instance, cadao, bananas, and palm oil from West Africa are exchanged for canned beef and cotton textiles. Tropical agricultural products are supplied to the mid-latitude countries from three sources. Most simply, wild products are collected by people who wander through forests and grasslands. Chicle was gathered in Middle America until the 1950's, and during World War II some rubber gathered in this way in the Amazon Basin. A second source of supply is the plantation, developed and controlled generally by out capital. The rubber plantations of Malaysia have for years supplied a large part of the world's natural rubber. Bananas from Middle America and coffee from Brazil are typical plantation products. A third source, less but increasingly important, is the farm of the small peasant-owner of the tropics. The cacao farmer of Ghana and the small rubber farmer of Malaysia are examples of peasant-owners. About one-third of Malaysia's rubber now comes from small acreages owned locally.

All the cacao of Ghana-constituting one third of the world's production-derives from small farms in the Ashanti hills. In much of southeastern Asia and much of Africa, large plantations operated by Europeans were long the rule, and such plantations are still dominant in world trade. In much of the development of African tropical crops, however, cooperative societies care for production and marketing. The cacao board in Ghana and the coffee board of the Chagga tribes near Mount Kilimanjaro are good examples. Marketing boards attempt to even out prices paid to farmers from year to year so that they are less violently affected by annual price changes. Price fluctuations are especially typical with regard to plantation products.

The marketing boards—a postwar phenomenon—are most highly developed in the Commonwealth areas of West Africa. The former French areas of West Africa are without marketing boards and are thus subject both to the effects of price changes and to the exactions of a series of middlemen. Trade which flows between mid-latitude continents is based to a smaller extent upon natural differences (for most mid-latitude lands have similar endowments) than upon differences in the human environment. Much of the bulk of the transatlantic trade, for instance, consists of wheat from the sparsely populated prairies of Canada. The grasslands of Europe are more densely populated and do not produce the surplus of wheat which they once did. Europe, however, cannot rely solely upon Canadian wheat.

The sparsely populated grasslands of the Southern Hemisphere, particularly those of Argentina and Australia, provide further sources of wheat and flour. In the same way, the Manchurian grassland once provided overcrowded Japan with wheat and soybeans. In general, we may say that sparsely populated countries export food and that densely populated countries either are self-sufficient or import food. Also, old established nations tend to import raw materials, while newer nations tend to export raw materials. Examples of these tendencies follow. Ghana, a young and sparsely populated country, exports bauxite, manganese, and cacao. Japan, an old and densely populated country, imports iron ore and coal, as well as about one fourth of her food requirements.

Indonesia, a new but densely populated (in the core area of Java and southern Sumatra) country, exports petroleum and industrial crops, such as tobacco and kapok, but imports foods, especially rice. Canada, a relatively old but sparsely populated land, imports bauxite and iron ore but exports wheat. It is difficult to find examples of old but sparsely populated countries, for most of the sparsely populated areas have been late in becoming separate political entities in the national sense.

Agriculture and world trade

Since farming is the most widely spread of the major occupations, it is not surprising that products of farm, ranch, and plantation should bulk large in world trade.

The agricultural situation is not a stagnant one, however, for agricultural products face notorious rises and falls in prices, which may over a period of time effect major changes in the productive landscape. Fluctuation in the prices paid for agricultural products may result from changes in technology or from changing production conditions.

The development of a plastic base for chewing gum in the mid-1950's reduced the market for chicle almost to zero. Synthetic rubber production has caused great concern in the areas growing natural rubber. That the growing of natural rubber has not disappeared is due to its superiority as a product in certain vital uses, as well as the continuing cost of the synthetic product. It is interesting to note that Brazil the parent country of the rubber tree-has set up a butadiene plant to produce its own synthetic rubber. A plant may be built within six or eight months, while rubber trees require 10 or 15 years of growth to begin yielding and take 25 years of growth before the optimum yield is forthcoming. In Brazil the time factor was obviously the overriding concern.

Climatic and labour advantages may some times tend to promote activation of tropical economies. For instance, olive cultivation in many Mediterranean areas is declining owing to competition from the increasing production of inexpensive vegetable oils in the labour-abundant tropical areas. The oil marketing boards in West Africa, by their wise buying and selling policies, are in good part responsible for this situation.

Japan is an interesting example of a nation dependent on world trade. Until the late 19th century, Japan was self-sufficient in food and almost without foreign trade. Now about one fourth of her food must - be imported. To pay for imports, dollars and sterling must be earned abroad through exports. Being poorly endowed with natural resources, Japan depends upon a well-trained and abundant supply of labour, which processes imported raw materials, and exports finished products of but higher value. The value added by manufacture earns the foreign exchange used to pay the food bill. The decade of the 1940's marked a change in Japan's foreign-trade policy. In the 1920's and 1930's, markets abroad were flooded with vast amounts of poor-quality but amazingly cheap merchandise. This kind of merchandise could be produced because Japan's labourers were rather poorly paid. In the 1950's a new policy became evident.

Relatively little cheap merchandise is now sold abroad, and in its place there are goods of higher quality. Locomotives for Thailand's railroads and buses for the roads of Argentina are examples of the new trend. Labour is better paid in postwar Japan and no longer can a cheap-labour economy prevail. The quality of Japanese goods has had to improve as Japan has felt the effects of market flooding and the competition of cheap goods from mainland China. England has even a greater degree of dependence on overseas trade than does Japan. Only 40 per cent of England's food is locally supplied. However, her marketing problem is less acute than is Japan's, for she has political ties, which result in economic benefits, with the Commonwealth nations. Japan-without an empire-must rely on a market created by the good will resulting from high-quality, relatively low-cost goods.

Western Europe and Anglo-America together generate three fifths of the world's 44 international trade, by value. Recent figures allot 44 per cent of this trade value to the former area and 18 per cent to the latter. The leading trading nations in western Europe are the United Kingdom, West Germany, and France. The United States and Canada alone constitute the Anglo-American unit.

Canada's major export item is paper, and her predominant import is machinery. The leading export item of the United States is vehicles, while more money is spent on the import of coffee than upon any other product. Only a quarter of her trade is with her neighbour Canada, though two thirds of Canada's trade is with the United States. Within the western European area the imports and exports of the leading nations follow somewhat different patterns. The exports of the United Kingdom are highly diverse and can only be characterized as miscellaneous. Her imports list is headed by meat products. Only 15 per cent of her total trade value is with neighbours.

Machinery accounts for more than 46 per cent of the value of West Germany's exports. Her leading import by value is grains, predominantly wheat from Canada. About two fifths of her trade is with neighbours. France's leading export is steel, which finds markets in the rest of Europe, in Algeria, and in the African areas allied to France politically. :.Her leading import is crude oil, mainly from the Middle East. She does only a quarter of her trade with neighbours. It is interesting to note that in the three nations here considered, the proportion of trade with neighbours increases as access to ocean traffic (insularity) decreases. Other factors, such as mineral resources, amount of agricultural land, and favourable trade agreements are also involved.

Around the Caribbean clusters a group of countries that have a high degree of involvement in world trade and yet are for the most part economically underdeveloped. The geographer Brian J. L. Berry calls these the "poor trading" countries. The leading import of 10 of these countries is textiles. Five of the 10 rely on coffee exports, three on sugar, one on bananas, and one on wood exports. It may be said of these 10 nations that they trade slightly processed vegetable raw materials for processed vegetable materials, predominantly cotton products. Somewhat less far down on the "poor trading" scale are those five nations which trade agricultural products for industrial goods. Vehicles are imported in exchange for cotton by Mexico, for sugar by Guyana, and for coffee by Colombia. The Dominican Republic exchanges its sugar for steel, while Panama exchanges bananas for a variety of items, more than half of which by value is machinery.

The remaining countries on the Caribbean export industrial raw materials and form a functional cluster of sorts. Venezuela exports crude oil, which accounts for more than half of her value of exports, and in return imports vehicles. Trinidad-Tobago and the Netherlands Antilles both import crude oil from their neighbour, Venezuela, and export refined petroleum. Surinam exchanges bauxite for machinery.

Exports

We may categorize nations by the nature and dominance of their exports. Some nations depend heavily on the export of one or more minerals for foreign exchange. Similarly, other nations rely upon the export of one or more agricultural products. Finally, many nations rely for foreign exchange upon the export of both mineral and agricultural products.

Nations relying upon the export of *one Major* mineral are relatively few. Four nations rely heavily on the export of petroleum. Venezuela provides 20 per cent of the world's petroleum; Saudi Arabia, 17 per cent; Kuwait, 10 per cent; and Iran, 14 per cent. Of these, only Iran processes a considerable amount before export. Guyana and Surinam each supply 14 per cent of the world's bauxite ore for making aluminum. Bolivia is almost solely dependent upon its tin export to the United States for its foreign exchange and the maintenance of its internal economy. Those nations relying upon the export of two or more minerals may be said to be somewhat less dependent upon fluctuations in the world market. Seldom will demand for more than one mineral change for the worse at any one time. Only three nations are in this category. Peru produces about 17 per cent of the world's silver, but its exports of copper and petroleum are relatively insignificant in world trade.

Chile produces 13 per cent of the world's copper, most of its natural nitrate, and a considerable amount of iron ore. (The steel mills at Sparrow's Point, Maryland, were constructed specifically for the processing of Chilean iron ore.) Before being partitioned in 1945, Korea produced 31 per cent of the world's graphite and 10 per cent of its tungsten.

Nations which rely heavily upon the export of agricultural products are fairly numerous. Those with a dominance of one-crop exports are eight in number. Coffee is exported by Colombia, with 10 per cent of the world's production, and by Ethiopia, with 3 per cent. Cotton is the main export of both Egypt and Sudan. The former produces half the world's total of long-staple cotton, and three quarters of her export value is accounted for by cotton. Sudan produces extra-long-staple cotton and exports it all. In normal times Burma exports rice, mainly to her neighbours India and Ceylon.

Pakistan is a heavy exporter of jute and produces about 90 per cent of the world's total. Honduras depends largely upon exports of bananas and is the leading producer of this fruit. Ireland exports cattle predominantly to Great Britain. Two nations rely heavily upon of varied agricultural products. New Zealand relies upon butter for 16 per cent of its export trade, upon cheese for an additional 5 per cent. Three quarters of the wool cut of New Zealand is exported, and meat exports are considerable. Seventeen of every 20 pounds of butter produced in New Zealand are exported, mainly to Great Britain. Another Southern Hemisphere nation, Argentina has great agricultural exports. Between 60 and 80 per cent of its hard flint corn is exported, mainly to northwestern Europe. Argentine wheat is harvested in November and December and hits the northwest European market when local supplies are low.

Most of the major product-exporting nations rely upon both agricultural and mineral exports. In general, they have correspondingly less economic fluctuation than the one-product exporting nations. Eight nations in Asia rely upon multiple resources for their export trade. Turkey produces 18 per cent of the world's chromite, most of which it exports, and its tobaccos are world famous. Its neighbour Iraq exports petroleum and is the world's 'leading supplier of dates, which come from the vast groves just north of the marshes along the head of the Persian Gulf. India exports cotton, predominantly to Great Britain, through the port of Bombay, which was built by the cotton trade. India's major mineral exports include sheet mica and manganese (9 per cent of world production). Its neighbour, Ceylon, produces for export 8 per cent of the world's graphite, as well as large quantities of tea and copra.

From within the area of southeast Asia are derived vast quantities of varied exports. Malaysia produces for export two fifths of the world's tin and about 31 per cent of its natural rubber. More varied are the exports of Indonesia, with 9 per cent of the world's tin production, 5 per cent of its bauxite mined, 20 per cent of its rubber, and important quantities of tobacco. To the north, Thailand produces 12 per cent of the world's tin and exports major quantities of rice and teak wood. Eastward, the Philippines are the world's largest producer of coconut - products (copra and coconut oil), and one fourth of all its arable land is planted in coconuts. Iron ore from Mindanao and sugar from most of the islands bulk large in Philippine exports. Three African nations have important mixed exports. Ghana produces for export more than 4 per cent of the world's manganese, while the Union of South Africa produces less than 9 per cent of the amount mined over the earth.

Ghana produces a tenth of the world's diamonds, while South Africa's share, though fluctuating considerably owing to sales by the Congo and Sierra Leone, is higher. Almost one-third of the world's gold is mined in South Africa, and much is exported to the United States. Wine and lobster tails complete the list of South Africa's major exports. Nigeria produces for export 6 per cent of the world's tin (from the Jos Plateau area) and major amounts of palm oil and peanuts.

Imports

Textiles constitute the major import item of 12 African and 11 Asian nations, as well as of Australia and New Zealand. Only one European nation, Denmark, and one nation in South America, Paraguay, count textiles as their leading import item. In return for textiles, nine African nations export vegetable foods or raw materials: cotton, coffee, cacao, fruit, and grains. Two others export animals or animal products, and one exports iron ore. Vehicle imports are paid for by the export of wine from Algeria and coffee from former French areas of West Africa. Egypt and Kenya-Uganda pay for their petroleum imports with cotton and coffee respectively. Five other African nations have imports in the miscellaneous category and pay for them with agricultural products. One with miscellaneous imports, Zambia, returns copper as a leading export. Libya, Mauritius, and Morocco are outstanding in the respect that both their major import and major export items are agricultural in nature. Libya exchanges hides for needed grains and Mauritius exchanges sugar for needed grains, while Morocco exchanges its grain surplus for imports of sugar.

The industrial backwardness of Africa is illustrated by the fact that most of the nations or trading units have an agricultural product as their main export item; only three of the nations export metals (copper and iron ore). Nine of the 11 Asian nations or trading units which import textiles export agricultural products as the leading items of export value. Items of export include rice for three areas, rubber for two, and copra, jute, fruit, and olive oil for one nation each. Hong Kong imports cheap textiles for its growing refugee population and exports high-quality textiles to Europe and Anglo-America. Cyprus pays for its textile imports with copper exports.

Five Asian nations import predominately raw or processed industrial materials and export agricultural produce. Taiwan exchanges its sugar for chemicals; India, its tea and jute for petroleum; Malaysia, rubber from its plantations for crude oil; Syria, grains for needed petroleum; and Turkey, its famous tobacco for machines. Two nations of Asia are agricultural in both leading export and leading import items: Ceylon exchanges tea from the Kandy highlands for rice from Burma and the Coromandel coast of India, and Israel sells fruit to buy grain. Iraq exports crude oil to buy vehicles, while Iran's crude oil exports pay for sugar. Agricultural products are the leading export items in 16 of the 23 Asian nations or trading units.

Nations with similar political interests often organize themselves into blocs in which currencies are freely convertible and in which considerable agreement on economic policy exists. We speak of Canada and the United States, together with their territories and possessions, as the "dollar area." The French Union is the "franc area," the British Commonwealth (minus Canada) is the "sterling area," and so forth. Occasionally, nations unite in agreements for specific purposes.

After World War II, concern for the poverty of much of the world prompted two major development program, the Point Four program and the Colombo Plan, as well as several minor ones. The Point Four program, now called AID, is supported by the United States government and is concerned with specific industrial, public health, population planning and agricultural projects. The Colombo Plan, supported in the main by Canada and the United States, applies particularly to areas in, or formerly affiliated with, the British Commonwealth. Within the framework of the United Nations, several aid organizations working toward similar goals include the World Bank and FAO—the Food and Agricultural Organization, headquartered in Rome. Aid is most frequently given in the form of trade items.

Textile industries are generally the first industries to appear when a country industrializes. This has been the case in the industrial revolutions in Europe, the United States, and Japan. If a nation is a major importer of textiles, it may be considered "industrially dependent." A map of those countries of the world for which textiles are the major import item by value is thus indicative, to some degree, of areas of industrial dependence. Agricultural exports bulk large in the foreign trade of nonindustrial areas; and as a nation industrializes, its proportion of exports which are agricultural in origin tends to decrease. As a result, a map of those nations whose leading export is derived from agriculture, animal husbandry, or fishing is also indicative of areas of industrial dependence.

Another indication of economic straits, if not precisely of "backwardness" is an unfavourable trade balance. A dozen countries in 1965 imported by value more than twice as much as they exported. The noncommunist portions of Korea and Vietnam received much of their imports as foreign aid, a good proportion of which consisted of military supplies. The reason for Panama's import imbalance is not clear, but that of Kenya may be related to its march toward independence. The remaining eight countries, which are either in or bordering the Mediterranean Sea, have chronic problems of trade imbalance.

Experience gained after 1947 through the Marshall Plan for European recovery resulted in the formation in 1952 of the European Coal and Steel Community, whose members were France, West Germany, Italy, Belgium, the Netherlands, and Luxembourg (the latter three are often called Benelux). By this arrangement the steel and coal resources of France and Germany were pooled under a supranational board. The plan worked well, and between the years 1953 and 1960 steel production doubled. The early success of the European Coal and Steel Community convinced the member nations that a broader organization might be feasible.

In 1957 the Common Market of Europe was formed, composed of the same "inner six" countries. Its performance pleased its members and created enough competition that Great Britain several times has petitioned for full membership. In 1961 Britain's Prime Minister stated, "The plain fact is that the formation and development of the Community has created, economically and politically, a situation to which we are compelled to react." Inasmuch as Britain is a member of the "outer seven" group called the European Free Trade Association, it is likely that Common Market membership will eventually be sought by others of the "seven," particularly by the Scandinavian nations.

Several countries are associate members, and many tariff concessions have been made to O.C.A.M. (Afro-Malagasy Common Organization), whose membership constitutes most of the newly independent African countries which were formerly French colonies. As a unit, the Common Market is now the leading importer of goods in the world and stands in second place in the export of manufactured goods. The member nations also cooperate in encouraging internal trade through tariff reductions and standardization of processing. Evidence of the Common Market's influence is the growth of at least 250 private merchandising and trade associations bearing "European" or "Common Market" names.

Foreign trade between Communist and non-Communist countries tends to be determined less by economic factors of supply and demand than by political factors. Exports to the non-Communist world from, Soviet-bloc countries in 1959 amounted to about 2.5 per cent of world trade. Economic benefits of such trade are often subordinate to political benefits. Two examples illustrate this. The U.S.S.R. in 1958 sold abroad more than 17,000 tons of tin at a low price. Possibly the Soviet area had an excess of tin, but it is generally thought that the market was flooded in order to embarrass financially the newly independent Republic of Malaya. Also in 1958 the mainland Chinese and private businessmen in Japan negotiated a trade agreement. The agreement was never put into effect, however, because the Chinese were not permitted to fly their flag over their Tokyo headquarters.

The Communists, not formally recognized by Japan's government, considered the flag more important than the economic benefits of renewed trade. In 1964, two thirds of the U.S.S.R.'s foreign trade was with Communist countries, and one fifth was with the industrial West. The Soviet version of the Common Market is the Council of Mutual Economic Assistance known in the West as COMECON. Though formed in 1949 as Stalin's answer to the Marshall Plan, it became defunct and was not revitalized until the summer of 1958. The Council's function is to forestall competition among parallel lines of industry within the satellite economies. Thus, it decreed, for instance, that East Germany (which produces one half of the machines imported by the U.S.S.R. and holds sixth place among industrial manufacturers) make trucks weighing less than four tons and that Czechoslovakia make heavier trucks. Standardization of parts has also been an aim of the Council; plans have been made, for example, to standardize machine tools for ball-bearing factories.

15
Manufacturing

Now manufacturing is a highly diversified and complicated operation, the production of each item involving many people, many different kinds of raw materials, usually some kind of machinery, and much transportation. Manufacturing is widely distributed, but most of it is concentrated in a few districts in which cities are engaged in manufacturing on a large scale and perform very special functions. The most extensive areas of manufacturing include northwestern Europe and northeastern United States and Canada, but other countries and other continents have smaller areas of manufacturing and all large cities have a considerable quantity of it.

The distribution of manufacturing can be measured in various ways. The way which involves determining the number of people engaged in manufacturing is easily applied because there is material available which indicates the number of people engaged in various occupations. Of course, the number of people engaged in a certain manufacturing process does not always indicate the relative importance of that process. In a flour mill, for example, most of the operations are automatic and require the services of only a few inspectors, adjusters, and maintenance people. Furthermore, if countries are compared according to the numbers, employed, those with relatively inefficient techniques are overrated.

Another method of determining the distribution of manufacturing is to measure the value added to products by the manufacturing process; the resulting value definition of distribution tends to underrate some industries with many employees, such as food processing. Still other methods of determining distribution are used, but these two are the most important and the most commonly used. Geographers are most concerned with the location and distribution of manufacturing and the study of related geographic qualities. Location factors are factors which tend to encourage the location of industries in certain cities or certain areas.

The most important factors which influence the distribution of manufacturing are raw materials, power, labour, transportation, and the market for the finished product. The importance of each of these varies with different industries and from time to time. The relative importance of the raw-material factor depends upon the weight and bulk of the raw material.

In the manufacture of cement, the raw-material factor is very important; the factory must be located close to the supply of shale and limestone from which the cement is made. At the other extreme, a watch factory need not be particularly concerned with the location of its sources of raw materials because the small quantities of metals and other materials needed in the manufacture of a watch can be easily transported thousands of miles, if necessary, at low cost.

The market factor is extremely complicated. Since any given weight of finished, products is more valuable than the same weight of raw materials, the cost factor is less significant. Cheap transportation thus serves to decrease the effect of the market factor on the location of industry. However, the location of the market is rather an important factor in the marketing of such heavy commodities as crude steel, many kinds of wood products, and cheap food products like grains.

As a commodity becomes more valuable per unit of weight, the market factor becomes more negligible. Thus, California can market citrus fruits in New York City in competition with Florida fruits, even though Florida is much nearer to the market. Of course, the location of the market is especially significant in relation to perishable commodities, including some vegetables and fruits and fresh milk.

The tax structure, which varies from city to city, from state to state, and from country to country, is likely to have an influence upon the location of certain industries. Some countries and areas have a very lenient tax structure and will, in fact, grant tax immunity to industries for 10 years or more after they establish themselves. Some areas have high income taxes, some none at all.

Not surprisingly, taxes are one of the things considered very carefully before a new factory or a branch of a factory is located in an area. Such a factor as climate is of some significance in the location of industries, but climate is perhaps less important than it was in previous years since air conditioning now makes it possible to produce and keep uniform the kind of climate most favourable for manufacturing processes. For many years, because cotton thread tends to break very easily under dry conditions, the cotton-textile industry was located only in regions with a normally high relative humidity. As a result, the Manchester area of England and the New England and southern Piedmont areas of the United States became centers of textile manufacturing. Since air conditioning involves additional cost, such humid regions still have a slight advantage.

Within the continental United States, location factors may make a crucial difference in the cost of producing a manufactured article which must be sold in competition with domestic and foreign articles of the same kind. Since each location factor is related to other location factors, the cost can be evaluated successfully only after all factors have been considered. Today there is a very intense study of location factors being carried on by experts.

Manufacturing regions

In the world today there are four major manufacturing regions which supply a very large proportion of the world's factory products. One region is in eastern North America and includes parts of the United States and Canada. Another, located in northwestern Europe, runs from southern Great Britain, across northern France into Germany and Czechoslovakia and includes Switzerland on the south and the Netherlands on the north.

The third region lies in the Russia. and includes the area around Moscow, the Donbas region of southern Russia and a number of somewhat more scattered areas in the Ural Mountains and Asiatic Russia. The fourth region includes eastern China and Japan. These four major regions have certain favourably location factors in common. For example, all have access to high-quality coal. Iron ore is either found within each region or readily available by water transportation.

All the regions have access to petroleum, although western Europe and Japan are deficient in this resource. Water power is available in all the regions and plays an important role in each. Except for some of the areas of the Russia, all of the manufacturing districts are fairly close to the sea. The variety of goods manufactured in these regions runs to many thousands of items. All regions have iron and steel industries, petroleum refineries, and metal-manufacturing plants for such diverse products as ships, automobiles, radios, and tools. Chemical industries produce fertilizers, drugs, and plastics; glass, cement, leather, and wood industries make up most of the remaining manufacture.

All the regions emphasize food processing-milling, baking, canning, freezing, and drying-and the manufacture of textiles and clothing. The tendency is for more and more food and clothing to be processed in factories and less to be processed in homes. The relative importance of these products varies widely, depending in part on location factors and in part on such other factors as local customs, average income, and purchasing power.

The relative importance of the different regions and countries is commonly expressed either in terms of the number of persons employed in manufacturing or in terms of value added by manufacturing. These data are approximate since the census methods vary somewhat from country to country.

The North American manufacturing region is, in many ways, the most efficient and highly developed of the four regions. It is especially noted for its large-scale development of machinery and such advanced techniques as the assembly line and very fine division of labour in manufacturing processes. This region has the advantage of water transportation in the Great Lakes and the St.. Lawrence Seaway and on rivers.

Numerous and well situated, the harbours along the Atlantic afford the manufacturers of the region access to the hinterland. The region is well served by railroads; few parts are more than 10 miles from a rail line. In the broadest sense, the hinterland of the region may be considered as including all of Canada, all of the United States, and most of Latin America. Internally, the North American manufacturing region is highly diversified and specialized. For example, heavy industry is concentrated in or near the coal fields or on navigable waterways; cotton textiles are in southern New England; automobiles and farm machinery are in southeastern Michigan. The following subdivisions are generally recognized.

1. New England is the oldest part of the North American manufacturing region. It has very few raw materials, no coal, iron ore, or petroleum, very few metals of any kind, and almost no industrial crops. New England contributes skill and specializes in light manufacturing, textiles, leather goods, hardware of all kinds including almost everything found in a hardware store, also small arms.

2. The New York metropolitan district is the most important single area of manufacturing in North America and employs more people than any other subregion. Light industries are important, especially those that require small space: food processing, clothing, printing, light metal goods, and refining of sugar, petroleum, and metals.

3. Southeastern Pennsylvania has both heavy and light industries. Originally it depended upon anthracite coal and local supplies of iron ore for its steel industry but today iron ore and coal are brought in by water to the tidewater mills. Oil refining, chemical industries, textiles including cotton, wool, and synthetic fibres are a part of the manufacturing complex. Philadelphia, Camden, Wilmington, Trenton, and Baltimore are the important ports and manufacturing cities of this region.

4. The Albany-Buffalo district is also called the Erie Canal belt because of the importance of this canal in the early days. This district is characterized by a high degree of specialization in different parts. In the eastern end, at Albany, Troy, and Cohoes, clothing manufacturing is important. Utica specializes in light

metal goods and textiles. Syracuse produces automobile parts, foundry products, ceramics, and shoes. Rochester is widely known for its optical goods and Buffalo, on the west end, has steel mills and is the principal flour-milling city of North America.

5. The southern Ontario district of Canada is in many respects similar to the Albany-Buffalo district. Coal is transported from the Appalachian region of the United States and iron ore comes in from the upper Great Lakes and also from the Labrador-Quebec region. Hamilton and Toronto have steel mills and Toronto manufactures clothing, electrical articles, and machinery and is also a meat-packing and a flour-milling center.

6. The Montreal-Ottawa district on the lower St. Lawrence River profits by large supplies of cheap hydroelectric energy. Paper and aluminum refining are very important, as are also textiles, clothing, flour milling, and sugar refining.

7. The Pittsburgh-Cleveland district is the most vital part of the manufacturing belt in North America, partly because of its central location, partly because of its large supplies of cheap, high-quality coal. It is dominantly a metallurgical center but there are also glass, ceramic, rubber, paint, machine tool, and electrical industries. It uses the upper Ohio River and its tributaries as well as the Great Lakes for water transportation.

8. Southeastern Michigan is noted mostly for automobile manufacturing. Nearly all of the automobiles made in North America come from this district, including Windsor on the Canadian side. Many of the automobile parts, however, are

produced in other regions. Farm machinery, furniture, and hardware are also important.

9. The Ohio-Indiana inland district including Cincinnati, Dayton, and Muncie, Indiana, is the only important manufacturing area not on the Great Lakes or on tidewater. It does have water transportation, however, by way of the Ohio River and has good rail connections to Pittsburgh. The industries are medium to light in character: machine tools, electrical machinery, radios, refrigerators, soap, shoes, clothing, and agricultural machinery.
10. The Chicago-Milwaukee district lies along the southern shores of Lake Michigan and via the Great Lakes has ready access to iron ore, coal, and limestone for the steel mills in the vicinity. Industries include food processing, flour milling and meat packing, agricultural implements, shoes, furniture, books, printing, and various types of metal fabrication.

The European manufacturing region includes parts of Britain, France, the Netherlands, Belgium, Germany, Czechoslovakia, Poland, and Switzerland and is connected with a few outlying areas in Italy, Spain, and Scandinavia. This region depends heavily upon coal, especially in Britain, northern France, and Germany.

All of the region is close to the sea and also benefits from excellent water transportation provided by rivers and canals. The products of the region are marketed in many countries but western Europe is its own best customer. The European region is the oldest and, in many ways, the most skilled in manufacturing, although it is only recently that it has developed the assembly-line technique which is so characteristic of the North American manufacturing region.

The European region has the special advantage of a very large pool of skilled labour. As a result, it not only exports manufactured products in large quantities but also actually exports skilled labourers to other regions. The European manufacturing region, like that of North America, is specialized by districts and extends from the central lowland of Scotland to northern Italy and western Poland.

Since parts of ten countries are involved, tariffs and trade agreements play an important role. In recent years, the tendency has been to relax trade barriers gradually, beginning with the Benelux (Belgium-Netherlands-Luxembourg) agreements and spreading to adjacent countries. An organization was formed, the European Economic Community (also called the Common Market), including France, Belgium, the Netherlands, Luxembourg, Italy, and West Germany. Later a somewhat similar rival group was formed, called the European Free Trade Association, including Great Britain, Norway, Sweden, Denmark, Portugal, Switzerland, and Austria. There are still strong trade barriers between these two blocs.

The chief subregions of the western European manufacturing region are as follows:

1. The Scottish Lowland district, with Glasgow on the Clyde River as a focal point, includes a great many industries. Iron and steel, shipbuilding, and textiles are the major ones. In the eastern part of the lowland near Edinburgh and to the north, lighter industries are more prominent, textiles, earthenware, and clothing.
2. The northeastern England district includes the coal field at Newcastle, near the sea. Iron ore is brought in from Sweden via the Norwegian port of Narvik. The iron and steel mills along the Tyne River supply the steel for a shipbuilding industry which rivals the Clyde River in this respect.

3. The Yorkshire district on the east side of the Pennine Chain is also based upon coal. The shallow coal has mostly been exhausted, the shaft mines are deep, and the coal is expensive to mine. Iron and steel are important, as also are woollen goods for which there are heavy imports of wool from Australia, New Zealand, and South Africa The metal industries include locomotives and machine tools.

4. The Lancashire district on the west side of the Pennine Chain has been the principal center for cotton textiles in Great Britain for many generations. Originally this industry was favoured by the moist climate, especially favourably for cotton spinning. Nowadays air conditioning makes it feasible to spin cotton any place. Manchester is the principal financial center of this district, and Liverpool is the main port. The principal textile centers are in outlying cities, Bolton, Oldham, and Stockport.

5. The Midland district to the south of pennines in the vicinity of Birmingham sometimes called the "black country" because of its numerous coal fields, coking-plants, and factories. Birmingham has good,rail connections and profits by its central location, but it has no water routes to the sea. Industries include iron and steel, hard ware,-, automobiles, motorcycles, bicycles, machine tools, radios, and ready-made clothing.

6. The southern Wales district in the vicinity of Cardiff and Swansea is based upon a good coal field and a location on the sea. Iron and steel are important but the outstanding industry of southern Wales is the refining of metals, especially tin, copper, and zinc, and the plating of thin sheets of steel with these metals. Ores, crude metals, petroleum are imported.

7. The London Metropolitan district is comparable to

New York in North America. London is in a good position to import raw materials, including chemicals from Germany and coal from various places. It has a large chemical and soap-making industry, printing, clothing, machine shops, and even me shipyards. Watchmaking, instruments, furniture, and glass are also important. A very large part of the manufactured goods consumed in the urban area.

8. The France-Belgium district occupies a long, narrow coal field from Calais on the English Channel through Liege in Belgium to the southern tip of the Netherlands. Good quality coal and iron ore from Lorraine in tern France form the basis for an iron and steel industry. This region also produces cotton, rayon, and woollen textiles, and clothing. It is well served with canals which connect with the lower Rhine River.

9. The Low Countries district along the coastal fringe of Belgium and the Netherlands is quite distinct from the districts which have coal fields. Imports from overseas areas, some of which were former colonies of Belgium and the Netherlands, supply raw materials for processing. Linens, lace, sugar and petroleum refining, rubber, copra, metals, chemicals, fertilizers, synthetic textiles, pottery, paper, and furniture are all important in this district.

10. The Ruhr district east of the Rhine and north of the Ruhr River is a small, compact district with at least 15 important industrial cities. It has the best coal field in western Europe and consequently has much heavy industry. Iron and steel mills supply the factories which turn out locomotives, boilers, railroad cars, munitions, and cutlery. Chemicals, textiles, and hardware are also important.

11. The Saar district is a small, strategically located area

on the northern border of Lorraine. It is close to iron ore, has good rail and canal connections, so that coal is available including some from local mines. Iron and steel and metal fabrication are the most important industries.

12. The Saxon Basin, in the vicinity of Leipzig and Dresden, has long been considered an important industrial district. It has large reserves of lignite, locally called brown coal, but good coking coal must be imported. The brown coal, however, forms the basis of chemical industries and is also used as a source of power. The Saxon district has long been noted for optical glass and instruments, such as microscopes, telescopes, and cameras. At Dresden, deposits of good clay form the basis for a porcelain and china industry. Leipzig is especially noted for chemicals, textiles, and printing.
13. The upper Silesian district centering around Gliwice includes good deposits of coking coal, ores of lead, zinc, and other metals, and also glass sands. Textiles, especially cotton, wool, and flax, are important. Heavy industry is, outstanding, comparable to that of the Ruhr in Germany and the Donbas in the Russia.
14. The upper Rhine district, including Strasbourg, Mannheim, Frankfurt, Stuttgart, and Zurich, is the poorest in terms of raw material of any of the European districts. But like the New Englanders in North America, the west Europeans developed a high degree of technical skill and do not require large amounts of raw materials to turn out their watches, light metal goods, textiles, and scientific instruments.

    Many of the factories are small and home industries still survive. The abundance of water power makes it possible to refine aluminum.

A number of outlying districts occur in western Europe, including central Sweden (15), which produces quantities of paper, pulp, wood products of all kinds, textiles, and small amounts of high-grade steel. In east-central France, Lyon (16) produces silk and cotton textiles and leather goods. In northern Italy, the Po Plain (17) is quite important for textiles, including cotton, silk, and rayon, and for automobiles. None of these outlying districts have good supplies of coal but as a partial compensation they have an abundance of fairly cheap hydroelectric energy.

The manufacturing regions of western Europe and North America are similar in a number of ways. Western Europe has good supplies of coal although in some fields it is expensive to mine. Petroleum is in short supply; only Poland, France, and Italy produce significant quantities. Iron ore is in good supply but many other industrial minerals are relatively scarce and additional amounts must be imported. Water for power and industrial use is adequate and transport facilities are very good. In western Europe no single country has a large, protected domestic market for its industrial goods.

The necessity for most countries to import large quantities of foodstuffs and other raw materials makes it necessary to find markets outside of Europe. The western European manufacturing region has shown more rapid industrial growth since World War II than has the North American region, but the latter still has a much larger per capita production. This growth has resulted in migration of workers from rural areas and from less industrialized countries into the regions of intensive manufacture.

The Russia. manufacturing region extends from Leningrad eastward through the Urals into parts of Asiatic Russia and southward to the Black Sea. Soviet manufacturing has been intensified recent years on the basis of techniques obtained from western Europe and North America, in part through the advice of many technical experts from these regions. Though the Soviet manufacturing system has been patterned on that of western Europe and North America, it has been modified by the Russian way of life. For example, the Russian manufacturing system has, put greater emphasis upon war materials of all kinds, certain transportation facilities including space craft, and agricultural machinery, and it has placed less emphasis upon what are generally referred to consumer goods.

The Soviet manufacturing region can described under three classes, the old, diversified districts especially Leningrad and cow; the heavy industry areas, the Donbas, the eastern Urals, and Kuzbas; and newly developed industrial districts including the Volga-Ural, the Karaganda, the Lake Baykal districts.

1. Leningrad, one of the oldest industrial districts, profited in the early days by its location on the Gulf of Finland with access via the North Sea to western Europe. Leningrad has declined in terms of its relative position in recent years. Metal refining, wood processing, paper, matches, and cotton textiles are the important industries.
2. The Moscow district has very few natural resources, but local supplies of low-grade coal, peat, and clay for ceramics are available. It has, however, a good transportation net of rails, rivers, canals, and pipelines.
3. The eastern Ukraine, including the Donbas, is by far the most important district for heavy industry.
4. The Volga-western Ural district, located on the middle Volga River and the west slope of the Ural Mountains, extends from Saratov through Kuibyshev

to Perm. One of the new industrial areas, it has access to metallic ores, oil and gas, potash, and timber from the north. Petrochemicals, synthetic rubber, synthetic fibers, fertilizer, and paper are produced.

5. The eastern Ural district, including Nizhniy Tagil, Sverdlovsk, and Magnitogorsk, has a great variety of minerals including magnetite. It gets oil and gas by pipeline from the Volga district and coking coal from Karaganda. Metals are important, as are chemicals, paper, and textiles.
6. The Karaganda district, including Temir-Tau and Tselinograd, has large supplies of good coking coal, part of which it exports to the other districts. Iron and steel and chemical industries are important.
7. The Kuzbas district, including Novosibirsk and Novokuznetsk, has good quality coal, easy to mine. Iron ore comes from the Urals, a long rail haul of about 1,000 miles. Iron and steel is the dominant industry.
8. The Lake Baykal district in the vicinity of Irkutsk, Bratsk, and Kansk has hydroelectric installations which provide power for metal, timber, and chemical industries.

A number of outlying industrial areas are being developed in the Soviet Union. Many of these are resource-oriented and are very highly specialized. These include the Fergana district in central Asiatic Russia. and small areas in eastern Siberia and Vladivostok and Khabarovsk.

In Japan, and to a limited extent in eastern China, manufacturing is growing rapidly in importance. The Japanese industrial belt is well developed and well known to the outside world, whereas less specific information is available for the Chinese region. The main Japanese manufacturing region is located in the southern and western portions of the country, extending from the vicinity of Tokyo westward for 800 miles to the west side of the island of Kysushu. It is a highly urbanized zone with a variety of industries on the west end, where most of the good coal is located and the iron and steel industry is well developed. The distribution of the other industries depends on location and transportation.

Especially noteworthy is the complementary relationship between Osaka and Kobe, and between Tokyo and Yokohama. Kobe is the deep-water port for Osaka and has heavy industries, while Osaka is a much larger city with light metal manufacturing; clothing factories are mostly in the outlying towns. Tokyo and Yokohama have much the same relationship and similar industries, although Tokyo is more a commercial than a manufacturing city. Nagoya has no deep, water port and is limited to the lighter industries—textiles, chemicals, machine tools; and ceramics. Japan's industries recovered rapidly after World War II and compete successfully, spite of tariff walls, for the world market in textiles, optical goods, electrical goods and wood products.

The manufacturing belt consists actually of many districts, separated by rugged terrain, tit connected by coastal waterways. Only a few areas are level enough for industry and these are intensively occupied.

1. The principal manufacturing district If Japan in terms of employment is the Kuant'ao area, also known as he Kuant'ao Plain, located on the margins if Tokyo Bay, including Tokyo, Yokohama, and about 25 smaller cities. Manufacturing highly diversified and includes foodstuffs, machinery, chemicals, printing, light metals, and optical goods. Shipbuilding and petroleum refining are important, especially in the Yokohama area.

2. The Nagoya district has been especially noted for its textiles: cotton, silk, and wool. It is also important for the production of aircraft, automobiles, chemicals, and pottery.
3. The Osaka district, including Kobe and Kyoto, is highly diversified. Formerly known for its textiles, in recent years there has been a shift to metal working and engineering industries, iron and steel, shipbuilding, machinery, aircraft, chemicals, oil refining, and food processing. All parts of this area have water transportation by way of canals.
4. The Kita-Kyushu industrial zone lies at the western entrance to the Inland Sea. It has a good supply of coal and specializes in heavy industry, iron and steel, fabricated metals, chemicals, glass products, and ship-building. Actually the number of factory workers here is not very great but this district has a large share of the heavy industry of Japan.
5. The Toyama district is one of the few industrial areas on the Sea of Japan. It has the advantage of large quantities of cheap hydroelectric energy and specializes in chemicals, especially electrochemical industries, paper, textiles, and drugs. Aside from these main areas there are hundreds of small industrial developments in various parts of Japan, including the Island of Hokkaido. Many of the products in these areas are for local consumption only.

Mainland China has great potential for industrial development. Large reserves of high-quality coal, iron ore, and various other metals are available. Agriculture provides additional raw materials, cotton, soy-beans, silk and other materials. The industrial development of China, however, has been very slow and is still small in comparison to the great population of the country.

Five-year plans have in the past aimed at greatly increasing heavy industry, particularly iron and steel, and only recently has more attention been given to consumer goods.

Prior to 1937 most of the industry of China was located in a few large cities near the coast, at Mukden, Tientsin, Shanghai, Nanking, and Canton. With the Japanese occupation of the coastal areas much of the industrial equipment was moved inland, particularly to Chungking, and at the same time the Japanese developed industrial sites in Manchuria. Following World War II many of the Japanese installations were dismantled; but they have been rebuilt to a considerable extent, and progress has been made in heavy industries. In addition, nuclear research has produced the hydrogen bomb. Even today, however, a large part of the basic needs of the people for consumer goods is taken care of by very small factories or handicraft manufacturing.

The South Manchuria district is in the province of Liaoning. Iron and steel industries based upon good local supplies of coal and iron ore are located in Aushan and Pench'i; metal fabrication is carried on at Mukden; chemicals, food processing (soy-beans), and textiles are produced in various cities.

Besides the four major regions, manufacturing is of great local importance in a number of places. Particularly in Brazil and India great strides have been made in recent years in converting from handicraft to modern manufacturing.

16

Ownership of the Land

Early in the occupation of an area the population density is generally fairly low. There are not many more people living on the land during the first stage of migratory agriculture than there were during the hunting and gathering stage. In the practice of migratory agriculture people go into an area of virgin forest, cut down the bushes and trees, and burn them to provide fertile ash and sunlight for growing two or three harvests before moving on. Since migratory agriculture makes a more reliable source of food readily available, the population increases gradually over what it was in the gathering stage. Also, the need for labour to care for the fields may induce the cultivators to have more children. In the practice of migratory agriculture land is considered to be available for use by all without restraint.

When the fertility of an area is exhausted after a year or two the cultivators move on, generally as a small clan of three or four families. In areas where the population density remains low it is possible that a clan will seldom return to lands it has once abandoned. In many areas the people "fill up the land," relatively speaking, and once-cleared fields that have since grown up in bushes and small trees are cut and burned again. Clans stop migrating in a to-and-fro drift, and settle down in villages of perhaps three or four clans which consider themselves to be part of a tribe. Fields are used, retired from use, and eventually used again in a system of "bush fallowing" developed under tribal rules.

Although the rules for the use of available land are generally simple when tribes are small and "bush fallowing" has just begun, it eventually becomes necessary for the tribal chief to make decisions as to proper use of the land. In general, the complexity of customs and laws connected with land tenure is directly related to duration of settlement; that is, the longer an area has been settled, the more complex its system of land tenure is likely to be. This rule is true of both tribal areas and more developed regions. It is evident that attitudes toward the use of a resource tend to vary with the amount of the resource available. Examples illustrating this tendency may be found in many areas of the world.

Among the pre-contact Tolowa Indians of California, anyone could fish freely along the ocean beach; along the much more restricted river banks, however, no one could fish outside of his allotted stretch of bank. Fishing in the ocean was communal and unrestricted among the Kiwai of New Guinea, but definite property rights were applied to their limited agricultural land. Most nomadic herding groups hold their pasture in common, and some observers state that the nomads may even regard land as being free, like the air. Kirghiz nomads of southern Siberia, for instance, are "communists" in summer when the pasture is abundant, but they become quite "individualistic" in their attitude when they descend to their winter quarters "Where grazing land is limited and they use both natural and artificial boundaries to mark off private property.

Among the Rwala Bedouins of Arabia, there are no rules about the ownership of territory or the inheritance of land, but there are elaborate rules governing the inheritance of camels and the use of wells. These examples illustrate man's attempt to divide the land and its resources in a manner considered both right and proper. Man's ideas about what constitutes a just division are complexly interwoven with other aspects of his life, as the following sections will show.

Tribal tenure

Ancient Hindu law states that he who clears a piece of land is the owner of it, and that sages who know the past declare a field to belong to him who cut away the wood or who cleared and tilled it. These paraphrases from the Hindu folk code, the *Law of Manu,* state the general tribal attitude of tenure through use, or use occupancy.

Since the clearing of woodland and brushland is an arduous task, it is not surprising that where tools are primitive and few, the labour of preparing a new field for cultivation carries with it fairly permanent rights to the produce of that field. Tribal land systems predominate in the cultivated but otherwise technologically primitive areas of southeast Asia and in Africa south of the Sahara. All tribal system, of landholding have two main aspects: the relation of the cultivating family to the tribe, and the ownership and labour allocation within the family.

Within a tribal family, the allotment of the products of the land and the assignment of work tasks differ considerably from area to area. In southern Africa the Zulu require the men to care for most tasks connected with the cattle. In much of central Africa there are cooperative work associations of men; rather than working individually on their own plots, the men work on each other's plots in rotation. Cooperation of this sort among women is much more rare. Where nomadic tribal groups travel through areas of settled agriculture, a mutually profitable interdependence develops between the nomads and the farmers. In Iran the nomadic tribes follow by custom a certain well-defined migration route, on which each subgroup has the traditional right to use specific areas.

Although the passage of the nomads may cause losses to be inflicted upon the crops in the settled areas, the farmers gain the advantages of having a plentiful supply of butter and meat at relatively low prices when the nomads pass through. Similarly, in West Africa the cattle-owning Fulani sell milk to settled farmers while their cattle are grazing on the stubble of harvested millet and other crops and leaving droppings which enrich the soil for the next crop.

In areas where cultivators operate under tribal tenure, a farmer's right to the original use of a piece of land derives from his clearing of the land with his chief's permission. His right to continued use of the land and retention of its harvest, called the *usufruct* of the land, depends upon his membership in the community; that is, he derives security of tenure from his village status rather than from his maintenance of the plot of land.

There is in tribal land systems a distinction drawn between the ownership of land by the chief as a result of his sovereignty and the ownership of land by the tribal member as a right of usufruct. In a sense, this distinction is a precedent for our present concept of the difference between the political sovereignty which is derived from actual land area and the right of a nation's citizens to use that land area in whatever way they deem proper.

In most of the tribal-tenure areas of Africa there are two ways of obtaining the right to occupy and use land. One of these ways is dependent upon a man's status as a member of his kinship group. If his kinship group occupies a piece of previously unused land, he has the right—by virtue of his membership in the kin group—to cultivate it. The second way of getting the right to use land devolves largely from the idea of political rights introduced into Africa as a result of European political control. If a chief is politically appointed, a tribal subject of that chief has the right of clearing the land. In the Nyakusa area of southwestern Tanzania, villages come into formal existence whenever a new chiefdom is proclaimed.

The members of a village locate their house plots according to the directions of the chief, and each man tends to cultivate that part of the land that is found immediately behind his house. Once the village has been established, villagers cl extend their cultivation to any land which is vacant—that is, land which is not being cultivated by another villager or lying in fallow for an other villager. If, because of death in his family or some other reason, a villager accumulates more land than he can cultivate, the chief may try to persuade him to give some of this land up to someone else. If the cultivator exercises his right of refusing do so, the headman generally has to accept his refusal.

Improvements in the land are the tribal justification for the right to use the land. The most permanent improvement made in land results from the planting of trees, and it is not surprising that land thus improved rarely changes hands. In the Nyakusa area it is common for a man's land and the trees which he planted on it to be inherited by his descendants. If the heirs have enough of their own land and are not interested in continuing cultivation of the inherited land, they nevertheless usually keep the more important trees. In some instances, bamboos, because of their value as a building material, may be assigned to the chief.

Trees or plants that have a very short growing period generally go with the land. Ownership of trees may cause a certain amount of friction. In the Sukuma area south of Lake Victoria, planted trees remain the property of the planter and his descendants even after they have abandoned the land on which the trees are planted. Later groups that occupy the area are not permitted to cut the trees or to use their products.

In most tribal areas a cultivator is left in undisturbed occupance of the land so long as he remains both socially and politically correct. Once he conforms to the tribal mores, he is able to remain in continued possession of the land: thus his sons may cultivate the plots and hold the rights to grazing, water, and trees and firewood. In many tribal areas, rights established in the past are not only valid today, but increasingly receive the vigorous protection of national law.

When one man in a tribe temporarily has land than he can use, it may become necessary for him to let someone else use the land. When the occupier is approached by friends, neighbours, or relatives offering a certain consideration in return for permission to use the land, a type of temporary tenancy arises in the tribal system. This type of tenancy will be mentioned later in the section on tenancy.

In the Shan plateau of Burma each plot of land belongs to the lineage that first cleared it. If the land is cleared a second time and the same lineage is not interested in using it, it may then be utilized by a second lineage either upon the payment of a token rent or simply with a formal act of "borrowing."

In some hoe-culture villages which have tribal mores regulating land use, the natural increase in population may lead to the founding of new settlements in virgin land by people splitting off from existing villages. An older village may thus serve as mother village to several colony villages established within a walking distance of about six or seven miles; these colonies rely upon the village for markets, festivals, and political guidance. Communal ownership of farmed land was the common pattern in the New World before the Spanish conquistadores instituted a system of private land ownership and large estates. In some isolated areas of the high Andes the old communal system has survived to the modern day; in other areas of Spanish America, such as the *ejido* areas of Mexico, attempts have been made to create a new communal system. Those American Indian groups who were nonagricultural for the most part did not have any conception of individual land ownership (though tribal hunting areas were zealously defended), until a communal system was imposed on them by virtue of their confinement to reservations.

One of the reasons for communal tenure is that it gives each operating or cultivating family an equal amount of land (and thus serves as a type of temporary freehold system). A second reason often cited for communal tenure is that it gives equal income to the operators of land, either because of their actual work or because of their membership in the community. The equal distribution of income among all the members of the community is the ideal of the major Communist countries.

According to Lewis Morgan, a 19th century American anthropologist, "primitive communism" was once the prevalent form of land ownership. The more we learn about the history of early peoples, however, the less we accept this often repeated claim. This claim was not original with Morgan, for Confucius believed that some form of primitive communism existed in China during the prehistoric past. Much of the Confucian system was concerned with trying to achieve a simplicity and order in human affairs like that which was believed to have existed during the ancient time of communistic land ownership.

Early systems

There evidently was a communal well-field system in ancient China (circa 1200 B.C.). The ideograph for "well" in Chinese resembles the crossed lines American children use in playing tic-tac-toe. When land was divided into square farms in dry northern China, it was assumed that each of eight farmers would cultivate his own plot and that the ninth, plot, the center of the lines, would be cultivated jointly by the eight farmers. One ninth of the total produce of the entire area was sent to the local lord or to the emperor.

As the population grew and the area of arable land increased, reapportionment took place on an egalitarian basis; still the yield due the emperor remained one ninth of the total. In Japan rice farming was engaged in from about 200 B.C. At first, since the population level was low and the marshy tracts were widely distributed, there apparently was no competition for land; as the population grew, however, and competition for land developed, land was divided according to a system imported from China in the 6th century A.D. An elaboration of the well-field system, this system, the *jori*, was a method of dividing land into squares, with each person allotted so much land within the surveyed area.

The idea of having a periodic division and redistribution of village land carried over from a somewhat earlier period in Japanese history. Although it is doubtful if any land was redistributed in this way in Japan after the 12th or 13th century, communal ownership of land and periodic redistribution of this sort was followed until 1899 on the Japanese-influenced island of Okinawa. The Japanese displacement of the Okinawan king ended the system, except on the isolated isle of Kudaka, where the system remains to the present day. Kudaka is a fishing village with a very limited amount of agricultural land.

The fields lie in various land-quality areas, and each family has a field from three to four feet wide and from 80 to 100 feet long in each of the areas. In modern Europe there are only scattered remnants (in England and Switzerland) of ancient communal land tenure. A semicommunal system known as the open-field system was, however, common in northern Europe in the Middle Ages, spreading over Germany (where it was known as the three-field system) in the 8th and 9th centuries, and lasting in England from Anglo-Saxon times to the 19th century.

The system flourished in northern Europe, where few tree crops existed, and where it was rather easy to redistribute land. Under this system a village community divided its land into several open fields, which served as grain fields, pasture and fallow in rotation. Within each cultivated field each landowner farmed a certain number of strips that were divided from each other by grassy ridges. A man's strips of land were both temporary and scattered. They were temporary because, after grain had been grown for two or three years in a field, the entire field was fallowed and then planted to grass; while his former farming area was thus in use, the farmer transferred his operations to another few strips of land in a large field newly plowed for grain.

The farmer's fields were scattered because, according to the egalitarian division made by the village, a man was 'entitled to' his share of the better land and had to take his proper portion of the worse land. On the fields which were devoted to pasture in the course of rotation, a man had the right to graze his cattle in common with everyone else's cattle. The grassy fields of pasture were generally referred to as the village commons. Each villager also had rights in the commonly owned and rather extensive wastelands beyond the arable area. However, when the demand for wool textiles coincided with the rural depopulation from the Black Plague, the large landowners of England enclosed much of the former common land for the purpose of grazing privately owned sheep and rendered the land unavailable for use by the village communities.

Later, as the countryside was being rapidly depopulated by the cities demand for labour to sustain the Industrial Revolution, the open-field, or three-field, system for all practical purposes disappeared in England (except for a small area near Laxton). On the continent of Europe a similar rise of industry and extension of cash and commercial crops had similar effects. Today the three-field system is almost entirely absent from the continent of Europe.

Socialist land tenure

In the modern world, communal ownership has reappeared in several socialist countries. Marxist theory states that the evils in economic life are the result of private ownership, and it declares that these evils will disappear if private ownership of productive facilities, including land, is abolished. Marx proposed collective ownership and operation of land, and collective distribution of the profits from land.

Collectivization was opposed by many farmers, particularly the more wealthy ones, the *kulaks*. Great starvation in the countryside (partly owing to mass slaughter of cattle by owners as collectivization began) was the initial result of the attempt to introduce communal tenure. The eventual achievement of collectivization probably owes a good deal to the previous existence of the *mir* system. In common use since the second half of the 18th century, the mir system had two aspects: first, the collective use of indivisible plots of land such as forests and meadows (a practice common also in northwestern Europe, particularly in

England and Germany); and second, the periodic reallotment of farmland based upon the growth in population and the changes in family structure that had taken place since the previous division of the land.

In some areas the division of land was proportional to the number of adult workers of both sexes, and in other places it was proportional only to the number of males in a family. In the decade after the Russian Revolution the Soviet planners experimented with State farms of extremely large size in order to test the possibilities of large-scale farm management. The relative success of the experimental state farms, or *sovkhozi*, hastened the planners' decision to collectivize the land in 1928, and in a few years all but the most isolated of some 25 million farmsteads were incorporated into some one of the 200,000 new collectives, most of which were based upon farm villages. Because the Russian peasant at that time was most commonly either illiterate or undereducated, as well as unfamiliar with modern technology, the collective farms were serviced by machine tractor stations.

All machines for plowing and harvesting were centrally located there and were rented by the collective farms. The tractor stations were an efficient allocation of scarce resources, as well as being a source of political control. During 30 years of collectivization the number of collective farms dropped considerably (from about 200,000 to 94,000) and the level of educational and technical training among the collective-farm population increased. Because of the fewer farms and the higher level of training, the machine tractor stations were abolished in 1948, and the machinery formerly rented by the collectives was permanently assigned to them.

For many decades a day on a Russian collective farm did not constitute a workday. The members worked in groups or brigades, and received credit according to the number of workday equivalents accumulated. The more advanced and technical jobs might be worth three or four workdays per day of actual work, while jobs of tedious hand labour (as done by slow-moving old women) were perhaps worth only half a workday equivalent.

When the harvest was in and the profits were known, they were divided in proportion to the accumulated workdays. Farmers were also allowed the usufruct of an acre or so of land near the village. In the spring of 1966 reforms in workday credits were instituted. There are long stretches of time on a wheat or rye collective farm when the farmers are not required to be in the fields, and the farmer can help his wife and children on the private holding. After World War II about a quarter of Russian farm production came from such privately held plots, although in area the plots accounted for only a tenth of the farmed land. Also, each family on a collective is allowed certain livestock-generally a cow, a few pigs, and chickens. If private production exceeds the needs of the family, the excess products may be sold in "free" markets in the cities, and the income thus gained does not belong to the collective but to the farmer.

During the first years of communism in China there was a gradual elimination of the landed gentry, or landlords, who had controlled a good deal of China's land. Although many of the landholdings of the gentry were less than 20 acres in size, it was decided to eliminate landlords as an economic class in order to satisfy the great demand for their highly productive land (which was, in general, much more highly productive than land in the West). From 1954 to 1956 it was the policy to amalgamate the peasants who had received land from this land reform into agricultural producers' cooperatives. The goal of this policy was to have one third of China's peasant households in these cooperatives (organizations less highly developed than collectives) by 1957. In 1958 a policy switch instituted the giant communes-great rural agglomerations of collectives aggregating as many as 100,000 people each. Although the communes achieved some increase in farm productivity, the increase was not nearly as much as had been hoped for. By 1960 there were indications that the drive to have all crops turned over to the government was slackening and that each family might be allowed to keep a small plot for vegetables, a pig, and a few chickens in the traditional manner of peasants. The effect of the Red Guards movement, 1966-1968, upon the communes is not known.

Land redistribution in the eastern European Communist countries has generally followed the pattern of collectivization. However, in many cases collectivization has not been so successful as it was in the U.S.S.R. because local conditions have differed somewhat. In India the common lands of the village have served in the past a far greater role than they do at present. The pressure of population has become so great that much of the land formerly held in common for fuel or grazing purposes has been turned gradually into cultivated land.

The use of the remaining common lands is somewhat restricted by the caste system- the right to use the common lands is based in many cases upon the amount of private land that a person owns, which ownership in turn may be a function of his caste membership. Before the advent of the Spanish, land tenure in the farming areas of the New World was predominantly communal. In a system both pre-Incan and pre-Aztec, lands were held in common by the *ayllu*, or village community. The old ways of cultivation also persisted after the founding of the Aztec empire in Mexico.

When the Spanish came to Mexico and Peru, they took possession both of vast areas of ayllu land and of the Indians who lived upon it. In the high intermont basins of the Andes they established large haciendas (which are still in existence) comprised of several ayllu each. Many farmers in the more remote or isolated ayllu still live in their preconquest fashion and give a certain percentage of the crop to their distant landlord. Those communities that are closer to the hacienda house are often directed by the landowner, or his agent, and may grow commercial crops in addition to those needed for community subsistence.

The hacienda, as a privately owned estate incorporating several communal ayllu, is an instance of the combination of two systems of land tenure. Advocates of land reform in Latin America want to abolish the haciendas and return the profits of the land to the *ayllus*. In 1940 the First International Conference on Indian Life supported the continuance of the ayllu, both because they perform a valuable social function and because their large areas are amenable to mechanization.

Though communal in theory, the *ejido* lands are most often held and cultivated in small areas by individual farmers and their families. However, in a few areas of Mexico the entire *ejido* area is cultivated communally some cash crop, with the profits from the crop being divided among the cultivators. Although most *ejidos* have not been mechanized, the farmers are not opposed to mechanization since any increase in income from mechanization would be to their advantage. The Mexican government does not tax the communal land of *ejidos*. Moreover, although the plots first allotted to *ejido* farmers were rather small and thus more suitable for subsistence crops than cash crops, the Mexican government has recently adopted a new policy of requiring more acreage per family in newly formed *ejidos*.

This increase in acreage will no doubt facilitate both the growing of cash crops and the mechanization of the land. It is ironic that most migratory Indians of North America, who never had any concrete idea of land ownership, should have been subjected to a communal tenure system by a government devoted to the achievement of freehold tenure for its voting citizens. Indian reservations maintain a form of communal land tenure in areas where land tenure was totally unknown in pre-Columbian days.

Americans often assume a *priori* that systems of communal tenure are inherently less efficient than freehold systems. Although it is true that communal systems tied to traditions and religious rites are often inefficient and unadaptable, the fact that communal systems are not inherently unworkable is illustrated by the success of communal tenure in Israel. The establishment there of collective farms, *kibbutzim* has gradually replaced a subsistence economy with a modern commercial agriculture in which citrus crops are emphasized.

During the early settlement of New England, land was often owned and operated on a communal basis by Puritan groups; later, the Shaker sect came to operate communal farms in New England. With the settlement of the more humid parts of the Middle West, numerous religious groups opened communal farming operations. One such group, the Amana colony in Iowa, diversified their operations after 1930 and now also engage in the manufacture of high-quality household appliances.

Within the United States communal tenure is still practiced by choice in the Hutterite farm settlements of North Dakota. The Hutterite sect holds its land in common and farms cash crops so efficiently that the freehold farmers in the neighbourhood have forced legislation prohibiting further acquisition of farmland by

Hutterite communities. The predominance of group agriculture, taking communal in its widest sense, is evident in the taxonomy of world agriculture developed by the geographer Leonard Zobler.

Tenant tenure

When the people who work the land do not own it, they are considered to be tenants. (Much of the land outside the socialist countries is cultivated by tenants.) In return for use of the land, a tenant may either pay a cash fee or give his landlord a share of the crop. The origins of tenancy are diverse and involve historical factors, rates of population growth, and matters of preference. Tenancy tends to modify the use of the land in those areas in which it is practiced. In many areas the widespread occurrence of tenancy has given rise to movements for land reform. The following section considers types of tenancy, origins of tenancy systems, and the relation of tenancy systems to land use and land reform.

Types of tenancy

Cash tenancy occurs mostly in advanced regions where commercial agriculture is dominant: Europe, the United States, Canada, Argentina, Australia, and Brazil. Cash tenancy is defined as a system under which a fixed amount of cash is paid to a landowner or a company for the use of land for farming. After the landowner has been paid the fixed amount, any profit derived from the land benefits the tenant.

Sharecropping (where landlord and tenant share fixed proportions of the crop, regardless of its amount) is much more common than cash tenancy. Under this system, if there is a bad year, the crop is often distributed in such a way as to help the tenant through. If the year is good, both landowner and tenant see profit in their share of the crop. The proportion of the crop received by the landlord depends on population density, competition among tenants for available land often permitting the landlord to demand a greater share.

The proportion taken by the landlord also depends on the amount of capital which he furnishes. For instance, if he furnishes the seed, fertilizer, and oxen or machinery for plowing, his share is far greater than it would be if the tenant provided them. In most areas where sharecropping prevails, the amount taken by the landlord is established by the economic law of competition and is regulated by custom. Carleton Coon reports that in one village near Tehran the yield of the grain is divided into five parts:

> ... one for the owner of the land, one for the water, one for the seed, one for the owner of the oxen, and one for the farmer who has done the work.
>
> The landlord owns the land, and also the water, and usually he is called upon to supply the seed as well, which gives him three fifths of the crop. The oxen are owned by only a few men in the village.... Most of the farmers, who own no ox, receive only a fifth of what they produce.

In almost all areas devoted to share-tenancy, there is a very complex system of proportions which depends on whether the tenant or the landlord provides the manure, machinery and implements, chemical fertilizers, aid and advice, threshing machinery, and marketing facilities. Because of this complexity, it is not possible to formulate a hard and fast rule for the division of shares that will be strictly applicable, even to a small area. Although the cacao areas of Ghana were mentioned previously as being dominated by freehold tenure, a Ghana farmer who own§ land which he does not operate often permits someone else to plant a cacao grove on his land. In such a case, one third of the crop goes to the farmer, one third to the landlord, and one third to the hired labourers from the north. If the tenant farmer is able to operate the land without hired labour, he takes two thirds of the crop. In a sharecropping scheme in Nigeria, the company which cleared the land and settled farmers upon it took two thirds of the produce, leaving one third of the crop for the settlers.

Eventually, the fact that each settler cultivated 24 acres and got the income from only eight acres proved a great deterrent to increased effort in the scheme. Where tenants are not competing for the land, landowners often find that they must make concessions to them. In some areas tenants acquire a right of priority, especially if they bring estate land into cultivation; the labour they put into the land brings them a sort of semipermanent tenancy, and they have the right of usufruct in the vines, trees, and fruit trees which they plant. In some areas of Iran the tenant may continue to operate the land as long as the roots of the alfalfa he has planted remain in the ground. Iranian landowners often discourage the planting of gardens by their tenants since tenants who have gardens tend to become more prosperous and to develop a more independent attitude.

One rather significant relationship between environment and land tenure exists i in Iran, where it has been noted that the land with the highest average annual rainfall is the land with the greatest percentage of large landed properties and consequent high rate of tenancy. As the rainfall declines, the amount of owner-operated land increases.

Tenancy may come about as a result of such factors as invasion, indebtedness, land hunger, and personal preference. Tenancy systems have often originated in a time of trouble caused by an invasion or a military clique's assumption of power. Military groups often do not appropriate the land to themselves, preferring to sponsor the consolidation of agriculture under a different and sometimes initially rational system. In the Philippines the Spanish conquerors claimed all land for the Crown but recognized that persons actually working the land had rights of usufruct. Land not in cultivation when the Spanish took over the Philippines was often given to army personnel and politicians, who became extremely wealthy when this land was later brought under cultivation.

The problem of land tenure in the Philippines continues to plague each succeeding national administration. The pattern of large landholdings in such Spanish-conquered areas as Middle and South America and the Philippines resembles the pattern established in southern Spain when the *reconquista*, or reconquest of southern Spain from the Moors, was completed in 1492—the year Columbus sailed for the New World. Since small, owner-operated holdings were predominant in both northern Spain and northern Portugal at that time (as they still are), these areas were not open to those desiring vast landholdings. As a result, those soldiers of Queen Isabella who were unsuccessful in gaining estates in southern Spain quickly turned to the New World for land.

Indebtedness to moneylenders has proved to be another reason for tenancy. A succession of two bad harvests may force land owning peasants to become so indebted to a city merchant that he can eventually take over their land by mortgage foreclosure. Much of the rural land in Syria is owned by Christian merchants in the towns; these merchants have never farmed the land which they obtained through moneylending practices that reduced the former owners to the status of tenants. In Burma the rice areas of the lower Irrawaddy were once owned by the Chettyar moneylender caste from Madras; however, when Burma gained its independence in 1948, the government dispossessed the Chettyars and returned the land to the tenants. In Indonesia the Agrarian Decree of 1870, issued by the Dutch, forbade the alienation of land to Europeans and Chinese but made no provision against the alienation of paddy land to the well-to-do Indonesians of the cities who regularly made loans against mortgages on rice land. Because of the high interest rates that these men generally charged borrowers frequently could not repay the loans and became tenants upon their formerly freehold land.

In India, the original zamindars, or tax collectors, leased out their interests, and the *zamindari* system gave rise to various levels of tenancy. As the population grew, those who leased from the zamindars leased in turn to others, until a long chain of tenants and rent receivers was formed. Proprietory rights in eastern Bengal were often held at seven or eight levels! In the process of land reform, the Indian government has reduced the amount of land held by zamindars from 43 per cent to 8 per cent of the arable areas.

Various levels of tenancy also exist in southern India, where temples own large landed properties acquired through gifts. The temples lease their land to rich tenants who, in turn, sublease it to cultivators. After he has made a cash payment to the temple in return for a lease of perhaps a dozen years, the first tenant leases or sublets the land annually at a rather large profit from shares in kind.

Tenancy has also originated in land hunger brought about by population increase. Population pressures often result in greater intensity of land use, under which a small area may give the same profit as a larger plot under less intense cultivation. In an area of intensified cultivation, a farmer who owns a certain amount of property may rent out a part of it, particularly if he has a social or ritual obligation to lease land to friends or relatives. In Ceylon tenure systems are quite often arrangements between relatives; in Japan before the 1948 land reform, much of the rented land was held by family members.

The social obligation to rent to family members is so strong that many of the colonists on newly reclaimed government land in Ceylon illegally and secretly rent out I d to relatives and other persons. Population pressure in the Philippines has resulted in subtenancy. An agreement is made between an owner and a tenant, called *inquilino,* who pays a certain cash. Although the inquilino owns the farm tools and animals and finances the farm operations, he often has share tenants under him to do the actual farming; such share tenants, called *kasamas,* amount essentially to agricultural labourers rather than tenants.

The sharecropping- form of tenancy is often preferred by rather wealthy tenants in the United States because it permits fluidity of farm operations and mobility of tenants. If a sharecropper's family labour supply decreases, he can operate a smaller farm; if his labour supply increases, he can rent a larger farm. In such areas as northern Spain and northern Italy the sharecropping system works out extremely well. In those areas the landlord is resident, takes an active interest in the manuring of the land and the selection of good seed, and supervises the farming. He also arranges for cooperative buying and marketing and is more intimately involved in the system than is the landlord who leases for money or the owner who comes to the country only at harvest time.

In the United States, most farmers desire, of course, to own the land they operate, but in many cases economic competition forces them to be tenants, at least in part. In his study of farming in the spring-wheat belt of the Dakotas and Montana, Walter Kollmorgen has found two important types of nonresident farmers—*side-walk* farmers and suitcase farmers. In regard to side-walk farmers (those who reside in town and commute to the farm, rather than living on the farmland which they operate), Kollmorgen states:

> Many informants in dryland wheat farming areas relate side-walk farming to absentee landownership.... [In Toole County, Montana] about 41 per cent of the side-walk farmers are full renters, as against about 27 per cent of the rural farmers. Conversely, rural farmers rate higher in full and part ownership.

In regard to suitcase farmers (farmers who live 30 miles or more from their operations), Kollmorgen says:

> Absentee ownership—largely by suitcase farmers themselves—and suitcase farming do seem to be closely related.... For the most part suitcase farmers must either buy land or find absentee owners from whom to rent. It is not popular in the County to rent land to suitcase farmers.

Sidewalk farmers and suitcase farmers both generally operate more than one farm, and their holdings are scattered for protection against the vagaries of drought, insect damage, and plant diseases.

The relation of land-use patterns and emphases to tenure systems is indirect and by no means always clear. In a cash crop economy there seems to be a tendency for tenant-operated farms to become larger as a result of the economies of scale connected generally with mechanization. In the areas of subsistence economy and a growing rural population, fragmented holdings and smaller units of holdings tend to be characteristic of tenant farming. For instance, in the Kikuyu area of East Africa tenants may rent from two or more landowners and thus have fragmented and scattered plots of land.

From the tenant's point of view, one drawback to the tenancy system is the fact that a tenant who is leaving the land is not ordinarily compensated for any improvements he may have made on the land; by and large, any improvements become the property of the landowner.

In areas where land is held on long-term lease (in Hawaii both urban residences and rural farmhouses are often held on a long-term lease), all improvements must remain on the land at the termination of the lease. In areas where many people are competing for tenant tenures, the new tenant may have to pay the landowner a certain premium for the privilege of cultivating the land; land held under such circumstances may be used rather roughly, without practices conducive to soil conservation. On the other hand, in areas where the tenant has the traditional right to pass on his holding to his son, the land is used in a more rational and soil-conserving manner.

In a study of tenancy systems in Malaya, L. A. P. Gosling contrasted operations on sharecropped land, owner-cultivated land, and cash-tenancy, or fixed-rent, land. He found that "... less than one percent of share-cropped land is planted in fruit crops and vegetables; but more than 7 percent of owner-cultivated land, and more than 13 of fixed-rent land is so utilized." He further noted that "... total time spent in weeding the crop—an indication of good farming—averages only 18 man-hours per crop year on share-rented land, as compared with 23 man-hours on owner-cultivated land, and 37 man-hours on fixed-rent land."

Thus where the fixed-rent system is predominant, the tenant engages in much more intensive cultivation; also, the tenant in the fixed-rent system puts more of his land in those subsistence crops, such as fruit crops and vegetables, which he can sometimes sell in the markets. In addition, Gosling found a significant difference in yields, noting that "yields on owner-cultivated land average 12 percent higher than yields on shared land; yields on fixed-rent land average 23 percent higher than yields on share-rented land."

Land reform

The transfer of land titles from landlord to tenant is of major importance in many of the overpopulated areas of the world, where the demand for reform has been strengthened by the increasingly short supply of land. Much land in the world is owned by persons who, *though large landholders, do little toward cultivating the land. There is a tendency in the Middle East and much of Latin America for the ownership of land to be the prime base of prestige in political life.

Prestige is derived in these areas not from the bank account resulting from profitable management of land but from the number of acres or square miles of land owned, regardless of its quality or degree of cultivation. For a long time, political power in both the Middle East and Latin America has been limited mostly to landowners and, although the situation is changing somewhat, land still determines prestige. Those who want land reform do not necessarily want complete equality; they may want only to achieve greater equality by giving easier access to land, by breaking monopolies of ownership, and by reducing rents.

A major program of land reform was carried out in Pakistan in 1959, when between three and a half and seven million acres of farmland were sold to two million landless tenant farmers, the 6,000 former landlords to be repaid over a period of 25 years. Whether it is feasible to have widespread freehold tenure on such a small scale will depend upon the ease with which the peasants may buy seed, fertilizer, bullocks, and equipment normally provided by the landlords. In much of the Near East, land ownership has traditionally been dominated by large absentee landlords; where peasant proprietors or small owners exist, they farm only a small portion of the arable area. In much of Iran, for instance, small owners were (until the reforms of the late 1960's) confined largely to mountain regions and less fertile regions.

Where small-owner villages are found side by side with villages of sharecropping peasants, the former are better cared for, just as the small-owner lands are better cultivated. In areas where crop failures are frequent, the small owner may suffer more than the sharecropping peasant during a bad year.

The political ecologist B. M. Russett found that land inequality and labour force in agriculture are the two most important variables in explaining the amount of domestic group violence within 44 countries. Restudy of the same data by H. R. Alker, Jr., indicated that land inequality and location of a country outside Europe explained 62 per cent of the amount of domestic group violence. Since many tenancy systems result from foreign conquest, or are associated with foreign rule, land reform represents one dimension of nascent nationalism and anti-colonialism. Political popularity can commonly be gained by the promise of land reform. Certainly few tenants or landless labourers would disagree with the view of Pliny that large estates destroy empires.

Freehold tenure, otherwise known as fee-simple tenure, generally implies the right to buy and sell land freely and to cultivate it as one pleases during ownership. Applicable to both subsistence farms and commercial farms, freehold tenure is the preferred and dominant method of land ownership in the English-speaking areas of the world; it also is the preferred and increasingly dominant form of ownership in much of western Europe, Japan, and India. Freehold tenure has stable historical roots in many areas, but as a system it is somewhat unstable owing to fluctuations in crop markets. It is hard to define freehold tenure rigidly, for many tenant situations approximate freehold tenure in practice and many so-called freehold situations differ little from tenancy.

For instance, in the Yoruba area of Nigeria serfs cultivate the land for masters who hold their land under either tribal or freehold tenure. So long as the serfs regularly supply a designated amount of produce, they are left free to farm as they please; and the produce demanded is scarcely more than a government would exact in taxes. On the other hand, in the Midwest section of the United States (especially in the dry-fringe prairie areas) many of the farms are "owned" by farmers who have large mortgages at banks.

There is a conviction in Anglo-Saxon areas that freehold tenure is the basis of stable political life. Presumably, the farmer owning his own land need not be dominated by other persons and can, therefore, vote freely and make political decisions of consequence without fear of economic reprisal. There is also a belief that freehold tenure is the most efficient method of operating farms. That freehold tenure often contributes to efficiency is beyond doubt, but (as will be made evident by the material below) efficiency is not always or necessarily the result of freehold tenure.

There are conditions under which fee-simple ownership encourages poverty and under which certain types of tenancy status promote more efficient use of the land and the obtaining of greater profits. Just as in tribal areas the right to permanent usufruct was granted to those who first cleared and tilled the land, so in more civilized areas (urban-based states) there were laws according to which persons who improved land for cultivation or reclaimed land from waste were given rights as permanent tenants which approximated the rights of freehold tenure. In ancient Greece rights were conceded to farmers in order to encourage cultivation. Also, it was long ago established as a principle of Moslem law that "whoever cultivates waste lands thereby acquires the property of them."

A state or political unit will often guarantee occupance of land that amounts in security of tenure to freehold rights. An illustration of such occupance is given by C. K. Meek:

> In the Dera-Ghazi-Khan district of the Punjab the person who clears jungle and brings land under cultivation is protected by statute. He cannot be ejected as long as he continues to cultivate; his occupancy right is heritable; he can cut down selfgrown timber for agricultural purposes; he can build houses, though if he vacates his holding he can remove only the materials he has paid for himself; he can sink a certain type of well; and he can sublet his holding temporarily but not permanently.

In medieval Persia, after the Mongol invasions, the state held that people who reclaimed dead land acquired thereby the rights of ownership and sale. On the other hand, even in areas of freehold tenure such as Britain, land may be appropriated by the state if a farmer is not using his holdings wisely; although the farmer maintains ownership of the land, it is operated under state control. Governmental control may serve to diminish the extent of freehold tenure. Under Turkish law, even though a farmer owns the soil and its produce, he does not own the subsoil. In Canada, mineral rights for mines and oil under the surface are

The fragmentation of land under freehold tenure is now common in most areas of subsistence cultivation, including a great deal of western Europe and much of eastern and southern Asia. Subdivision or fragmentation of property was also practiced in early times. Among the American Indians, the Tolowa in California divided river banks into individually owned fishing grounds and possessed individual tracts for the hunting of deer.

The Nevada Washo Indians individually owned clumps of pine-nut trees and passed them along to their children. Likewise, the saying of the Yoruba in Nigeria that "a tenant should not raise his eyes upward" reflects the practice of property fragmentation. Among the Yoruba, the person who plants a tree or digs a well generally owns it always, even if he leaves the plot or village. Where land is regarded as the gift of nature, property rights do not apply to it until there is a capital improvement made on it through the expenditure of labour.

In the Middle East the system which allows one man to own the trees on another man's land may have originated in a very early period, possibly in the period when Moslem law was first applied. In Islamic areas it is the general rule that a man's widow gets one eighth of his holdings while his children get the remainder, each son receiving two shares for each daughter's one share. Generally, these shares remain undivided and the fruit of any inherited trees continues to be shared according to the original proportions.

In some Near Eastern areas there has been excessive subdivision of trees, with the result that plantations of date palms are rendered almost valueless because so many heirs own them. When there are too many heirs, no one person has enough trees to take responsibility for maintaining the plantings. Likewise, the substitution of better varieties of palms may be inhibited by the joint ownership of the palms to be replaced, perhaps even of the single palm to be replaced. In some places (Iraq, for instance) feuds are carried on not only by shooting people but by firing bullets Into an enemy's tree.

Freehold tenure is the dominant form of landholding in Japan and South Korea. Before the land reforms of 1948, small landlords (each with less than 25 acres of cultivated land) represented 90 per cent of the landlord class and controlled 55 per cent of all land. Although the peasants of Japan and Korea still do not own all the land they cultivate, the reforms of 1948 broadened the ownership base considerably. Since land lying within the boundaries of the house lot has traditionally not been taxed, over the centuries it has become some of the most intensively used land in Japan. In India there are two major categories of private ownership. So-called perfect ownership entitles holders of land to rights in the village common that are proportional to the size of their private holdings. The second type of ownership does not entitle the landowner to use of village common land. There is also a rapidly disappearing type of ownership known as *zamindari* ownership.

The zamindars (or landowners) are descendants of persons who, employed by the British in the 18th and early 19th centuries to collect revenue from the tillers of land, came to have proprietary rights over the land they administered. The *zamindari* system is gradually being replaced by *ryotwari* tenure, under which a farmer owns his land in fee simple and pays taxes directly to the state.

Carleton Coon has written of the Khunik villages in Iran, in which the villagers are both owners and cultivators and have a general air of prosperity. These villagers raise saffron as a cash crop and are able to keep a large proportion of their total income. A fifth of their income goes to the tax collector, a tenth to a local mosque, and perhaps another fifth to a family charity in a nearby city. The fact that any improvements he may make remain his own property encourages his exercising foresight and initiative. Much *mulk* land is held only on the basis of tradition, there having been no registration of ownership; in some areas whatever records once existed have been lost.

Absolute free hold ownership of land is comparatively rare; even the large tracts of absentee land lords may be held under the more common type of tenure, called *miri*. Although *miri* theoretically belongs to the state, most farmers on *miri* land are for practical purposes freeholders. Sometimes miri land is leased only temporarily, but it is often leased in perpetuity, under qualifying conditions. In Africa most of the outright freehold tenure is found in the Union of South Africa and in some areas of West Africa. In the freehold areas of South Africa most oi the work is done by black African contract labour; however, some freehold areas near the larger cities are becoming mechanized In West Africa the most successful small holders are the Ashanti cacao farmers of Ghana. Among the Ashanti each farmer generally owns from 12 to 20 acres and works them with the aid of his family and migrant labour coming from the Moslem north. In Uganda the British established a freehold tenure system which modified the political and other rights of the chiefs. Also, the rights of peasants on estates were defined and regulated by law.

When a subsistence economy was dominant, Uganda women provided most of the labour. With the introduction of cash crops, the men have become active as cultivators of the economic crops, while the women continue to produce the food for subsistence. It is almost impossible to generalize about freehold tenure in Europe. Freehold areas abut areas maintaining various tenancy systems. In southern Italy and Sicily the great estates or *latifundi*, extensively rather than intensively cultivated, lie side by side with small plots characterized by intensive cultivation. Northern Portugal is a land of small freeholdings and a high population density; southern Portugal has large estates and a low density of population. In France of the farmers own the land they operate. Freehold tenure is becoming somewhat more common in Latin America. For example, immigrants into Brazil from Italy, Germany, and Japan have settled in southern Brazil and have made a commercial success of freehold tenure in a country where small-scale freehold tenure has traditionally been associated with poor subsistence farming.

Unrestricted individual ownership of every acre of land in a country would make society unworkable. Since, in every country, certain uses of the land are thought to be important for the community as a whole, ownership is often lodged in a government or a corporation of some sort. Land not privately owned may be owned or operated by various levels of government, by legal corporations, or by religious institutions. Among governmental types of ownership, we can distinguish four: national, state, county, and city ownership.

National ownership of land results most frequently from conquest and less frequently from changes in government. Much of the land in the Middle East, North America, and most Latin American land became government property through conquest; in Latin America the land was soon parcelled out by the governments to private individuals. When the United States became independent of Great Britain, land belonging to the Crown was automatically considered to be public domain of the new country.

In many British colonies and countries of the Commonwealth, Crown land is the necessary step between tribal or communal ownership on the one hand

and private ownership on the other. In New Zealand the Maori conceded many of their lands to the Crown in return for a monetary consideration.

The assumption in English law that the rights of the freeholder are subordinate to the paramount powers of the Crown is the origin of the doctrine of eminent domain, whereby private land can be taken for public purposes upon the payment of adequate recompense. The greatest degree of national ownership in the United States is found in the western part of the country, where national parks and national forest lands occupy a large proportion of the area. Federal ownership also applies to canals, many of the new highways, and all of the air space. In the United States the national government has tended to reduce its ownership of agricultural land, while maintaining its control of forest land. Unlike Turkey and Canada, the United States allows freehold owners to hold subsoil mineral rights. Much of the public domain obtained through the Louisiana Purchase was distributed to homesteaders in the period from 1865 to 1890.

In the disposal of the public domain, one square mile section of each township was kept out of private ownership for the purpose of financing schools. (This reserving of a portion of land for the common weal resembles the well-field system of ancient China described previously.)

The ultimate origin of freehold land in the public domain is apparent to anyone buying land in the United States, for each deed traces the ownership of the land back to Indian treaties, or to original grants of land by English, French, or Spanish kings. In such commercial economies as those found in Australia and the United States, pasture is most often owned and leased out by the government. The Taylor Grazing Act of 1934 codified and regulated United States policy on the use of government-owned pasture. In the widely occurring pastoral subsistence economies of Eurasia and Africa, pasture is most often communally owned.

National ownership of land results when the national government acts as middleman in land-reform transactions. During most processes of land reform, the national government buys the land from landowners and continues to own it until it is distributed to farmers in some manner. The new owners pay the government for the land over a period of time, generally 20 or 25 years, and the government reimburses the former owners for the expropriated land. Since modern governments sponsor research of various kinds, they may own land for the purpose of agricultural experimentation. Most governments do not hold a great amount of land for this purpose, but the large state farms, *sovkhozi,* of the U.S.S.R. are an exception.

Political units smaller than the nation often own land in their own right. In the United States, counties and states own forests and lands such as gravel pits for the repair of roads and highways. Since 1953 each ocean-bordering state has also had control of its offshore oil deposits. Most urban land is privately owned, but the city as a unit may own the parks, the city government buildings, and various areas where machinery is maintained for the upkeep of urban facilities and streets. In recent times some cities have purchased recreational areas beyond the city limits; for example, the Denver Mountain Parks area along the Front Range of the Rockies is about a dozen miles west of Denver. In New England certain towns now own property, in the countryside for industrial development and recreational use-property which was once owned by a small number of individuals. Ownership by a local political unit represents a return to a kind of communal tenure.

Corporate ownership of land

Ownership of land may be lodged in temporary corporations (such as the privately supported "land-gift" Bhoodan movement in India) and in legal corporations up for the specific purpose of operating areas as cemeteries and plantations or articular large-scale agricultural projects. Neither of these types of corporations is governmental in nature.

The Bhoodan movement was started by Vinoba Bhave, a disciple of Gandhi, who travelled around persuading landowners to give a portion of their land to be redistributed. Legal corporations in the United States and Canada way own cemeteries in order that persons belonging to certain organizations may be buried there; a corporation is able to provide a cemetery with "perpetual care." A plantation is usually dominated by a single cash crop and is, as a result, sometimes referred to as a farm factory. It is usually located in the tropics, where it has been initiated by a mid-latitude enterprise in response to demand for products not producible—or producible only at great cost—in nontropical areas. Since plantations both within the tropics and on their fringes have generally been associated with foreign capital and operation, they have often inspired political opposition.

However, when plantations have been established by production development boards under local governments (southern Nigeria is an example), political opposition has been quite limited. Production may involve either a local crop or an introduced crop like rubber in Malaysia or cacao in West Africa. Despite some instances of local regulation, capital and control are essentially foreign; and often the labour supply is foreign as well. (For example, many of the rubber-tappers in Malaysia are Tamils originally recruited in southern India.) Less than 1 per cent of the land area of the tropics is devoted to plantation agriculture, but plantations provide most of the tropical cash crops. Company plantations are generally much more extensive than private plantations; companies, for example, own large rubber plantations in Malaysia, cacao plantations on the islands off the Guinea coast, and banana plantations in tropical America. Such company-owned plantations seldom cause trouble except where they keep land out of cultivation. They may be responsible for certain undesirable social consequences, but they remain productive investments.

The large cotton plantations in the Gezira area of Sudan are unique examples of cooperation between a government, private corporations, and peasant tenants. The government of the Sudan built a dam for the storage of irrigation water and canals for its distribution. Two companies, Kassala Cotton Company, Ltd., and Sudan Plantations Syndicate, Ltd., held concessions to market the crops and gave technical advice and assistance to the cultivators. The cultivators got 40 per cent of the profit from cotton sales, the remainder of the profit being shared by the two companies and the government.

When the plantation system was nationalized in 1950 (upon Sudan's attaining its independence), the Sudan government assumed the role of owner and became the middleman in transactions concerning land management and cotton sales. The government has maintained the former provisions against sales of land to merchants, absentee ownership, and fragmentation of holdings.

Land owned and operated by religious institutions has been common at various places and times. In the Western tradition, monasteries with attached farms are familiar to everyone. In the 12th century the Japanese government supported Buddhist temples directly. Later any land dedicated to temples and developed either by them or by farmers assigned to them was held to be exempt from taxes. Under the impetus offered by this system, land reclamation flourished in areas belonging to the temples, while the revenues of the central government declined and the nation lapsed gradually into feudalism. In many areas where expanding population is exerting pressure upon a limited amount of cultivated land, the government often encourages the development of wasteland by giving remission from taxes for a period of years. Frequently, the reclamation of wasteland is done under the direction of a religious institution. Since the remission of taxes amounts to a subsidy that is supported by owners of land subject to taxation, it has been opposed in many areas.

17
Immigration and Asylum

Some are born stateless; some become stateless; some: have statelessness thrust upon them. The Independent Commission's analysis followed traditional lines, noting first the legal dimension of the problem:

> ... the faulty expressions of national legislations on the question of nationality and the principle by which the law of cacti State fixes the acquisition, attribution or loss of nationality.

Some States opt for the principle of birth within territorial jurisdiction as the necessary and sufficient condition for national status, while others also require an element of filiation, or descent from a citizen. Variations on these themes, and the inherent incompatibility of many permutations, frequently entail no entitlement to citizenship under municipal law, no matter what factors otherwise link the individual and the State. But even acquired citizenship may be lost, for example, because of overseas residence, or as a consequence of territorial transfers between States, because of punishment for political or criminal reasons, or, for women, in particular, because of marriage to a foreign national. The Independent Commission rightly noted, however, that the legal aspect was only one dimension. Others were stateless because of the effects of migration or coerced or involuntary displacements. The terminology is often confusing, the discourse moving from de jure statelessness to de facto statelessness, while trying both to accommodate and to exclude refugees from the delimited field. The core issue is that of being without protection, from which a host of other disadvantages flow, such as restrictions on freedom of movement, persistent uncertainty in dealings with State authority, transmission of statelessness from generation to generation, lack of identity or travel documents, and the absence of a sense of belonging. In practice, concluded the Independent Commission, 'the stateless are less protected than refugees."

Forty years earlier, in 1948, the United Nations Economic and Social Council (ECOSOC) was even then talking of the need to adopt interim measures to afford protection to stateless persons. but It was refugees they had mostly in mind. A study commissioned by ECOSOC and published the following year, locked statlessness firmly into traditional patterns of international law analysis. A limited phenomenon for most of history, statelessness and refugee flows emerged dramatically during the twentieth century. The product of territorial reshuffling and political and social crisis, periodically accompanied by lack of co-ordination and consensus on nationality rules. The United Nations Study of Statelessness in fact paid comparatively little attention to the non-refugee stateless, while expending considerable energy on delimiting categories; it distinction was thus attempted between refugees, who might also lie stateless de jure or de facto; and stateless persons, who might be stateless *de jure or de facto*. The Study was also careful to exclude from its ambit 'refugees' who were not stateless, such as internally displaced persons, as in Greece, or Germans, Indians, and Pakistanis.'

What was needed first was to improve the status of 'stateless persons', and secondly, to eliminate statelessness ' For receiving countries, such persons were an anomaly, possessing 'no definite legal status and without protection.' For the stateless themselves, their abnormal position might reduce their 'social value', and destroy 'self-confidence'. In the organized world of the late 1940s, status was what wits required together with international agreement on freedom of movement residence and settlement civil status, the exercise or trades or professions, education, relief, and social security. Most important of all, they needed protection, and that wits best achieved by providing identity documents consultor services, and legal status. So, ironically, the better protection of stateless persons was to be secured by offering them the status of stateless persons. At the same time, the phenomenon of technical, as opposed to refugee, statelessness, was to be eliminated by appropriate agreement; harmonization of nationality legislation; restrictions on deprivation of nationality as a penalty; better regulation of territorial settlements; and reduction of existing numbers through the facilitation of naturalization. ECOSOC decided to establish a committee to work on the definitions, and work out the solutions.

The ad hoc Committee on Statelessness and Related Problems met twice in New York in 1950. At the very beginning, the French representative, Mr. Rain sought to isolate two distinct issues: 'the status of refugees and the problems related to the legal status of stateless persons.' In his view, a draft convention on the first was urgently required, but the elimination of statelessness was basically different: '....rather a continuing concern of the world community than an acute situation which required immediate remedial measures,' In his view, there were two categories of stateless persons: those who were also refugees, and those who were not. Almost all refugees were in need hence the special urgency of their plight; but the same could not be said for non-refugee stateless persons. no United States representative, Mr. Henkin, agreed, and pressed for separate treatment; the draft convention should be limited to refugees, and '...should not be based upon a confusion between the humanitarian problem of the refugees and the primarily legal problem of stateless persons.' The problem of stateless nonrefugees should be kept aside, 'especially since there were doubtless stateless persons who were in no need of protection by the United Nations.

The British and Chinese representatives seemed less sure. The British noted similarities, and the need for similar entitlements; the Chinese recalled that, like refugees, stateless persons also lacked legal protection.

In the event, it is hardly surprising that the Ad hoc Committee decided to focus upon the refugee, strictly so called. It its report to ECOSOC, the Committee offered the text of a draft refugee convention, but simply proposed an additional protocol on stateless persons, under which States might agree to extent the application of the refugee convention, mutatis mutandis, to stateless persons to whom it did not apply, In August 1950, ECOSOC returned the draft convention to the Ad hoc Committee for further review, prior to its being considered by the General Assembly, and finalised the Preamble and refugee definition. At the same time, it welcomed the fact that the International Law Commission was to take up the subject of nationality, and invited States meanwhile to do what they could to reduce or eliminate cases of statelessness. In December 1950, the General Assembly decided to convene a Conference of Plenipotentiaries to complete the draft convention relating to the status of refugees, and the draft protocol on stateless persons," Meeting in July 1951 in Geneva, the Conference adopted and opened for signature the 1951 Convention, but decided to take no decision on the draft protocol, referring it back for more detailed study. Three years later, at a United Nations Conference in New York, the 1954 Convention relating to the Status of Stateless Persons was adopted, an independent convention being preferred to the draft protocol initially proposed by the Ad hoc Committee in 1950. Article I defines a stateless person in purely formal terms; 'the term stateless person means a person who is not considered a national by any State under the operation of its law.' Others among the unprotected are for the most part beyond the scope of the Convention, save that the Conference recommended that 'each Contracting State, when it recognizes as valid the reasons for which a person has renounced the protection of the state of which he is a national, considers sympathetically the possibility of according to that person the settlement which the Convention accords to stateless persons'.

The 1954 Convention's definition of its beneficiaries is essentially legalistic, conferring status on a relatively circumscribed class. The 1951 Refugee Convention certainly contemplated coverage for some categories of stateless refugees. but even with the above mentioned recommendation, the majority of stateless persons remain likely unprotected. Even those who are covered by the 1954 Convention are not as well off as refugees, having no equivalent guarantees against penalization for illegal entry, and no monitoring body like the UNHCR to which to turn for support against State authorities. Similarly, very limited protection is accorded in time of war or other armed conflict; the 1949 Fourth Geneva Convention on the Protection of Civilian Persons cites only the case of refugees: the Detaining Power is not to treat as enemy aliens, exclusively on the basis of their nationality, refugees who do not in fact enjoy the protection of any government.

The limited purposes of the 1954 Convention are worth recalling: to regulate and improve the status of stateless persons and, within the context of the purposes of the United Nations, to assure stateless persons the widest possible exercise of fundamental rights and freedoms. The class must he defined, and the rights, guarantees, and standards of treatment set out. The elimination of statelessness through the co-ordination and harmonisation of national laws was to require further international cooperation.

In 1959, after some preliminary work by the International Law Commission, a United Nations Conference convened in Geneva to consider a convention for the elimination or reduction of future statelessness. Fundamental differences were revealed between States which favoured the principle of jus soli, and those which opted for jus sanguinis. Endorsement of the former would have stopped many instances of original statelessness at source, but agreement was missing, and the compromise reached when the Conference reconvened in 1961 combined elements of both principles. Equally divisive was the issue of deprivation of nationality, a facility defended by many States as essential to their vital interests. In principle, deprivation resulting in statelessness is now prohibited by article 8, but subject to a variety of exceptions, including misrepresentation of fraud in acquisition, or disloyalty. Deprivation of nationality on racial, ethnic, religious ,or political grounds, however, is prohibited without exception.

As Weis remarks, one of the most significant elements In the 1961 Convention, is the fact that it imposes positive obligations on States to grant nationality in certain circumstances, in contrast with the essentially negative obligations contained in the 1930 Hague Convention. The 1961 Convention also attempts to settle a variety of incidental problems, such as the nationality of founlings (article 2: continuing the principle of jus soli already established in the 1930 Hague Convention); and of those born on board ships or aircraft (article 3). It seeks to minimise the possibility of loss of nationality resulting in statelessness on the occasion of change of ci 'il status (article 5 and 6), and even in the case of voluntary acts of the individual, such as renunciation (article 7). Finally, the Conference resolved 'that persons who are stateless defacto should as far as possible be treated as stateless de jure, to enable them to acquire an effective nationality'. As Weis has observed, this distinction will often be difficult to draw in practice, and he prefers the category of 'unprotected persons'.

At one time, the International Law Commission had favoured the idea of both a protecting agency for stateless persons, and a tribunal lo decide upon (heir claims. Neither suggestion found much favour with States, which opted instead for the establishment of a body within the framework of the United Nations, 'to which a person claiming the benefit of (the) Convention may apply for the examination of (the) claim and for assistance in presenting it to the appropriate authority' (section 11). When the Convention eventually came into force in 1975, the General Assembly requested the Office of the United Nations High Commissioner for Refugees to undertake the functions described in article 11, on a provisional basis.

Important as are the principles set out in these instruments, what finally counts is State practice. The 1954 Convention and the 1961 Convention have attracted few ratifications: at I January 1989, thirty-six and fifteen respectively; and the actions and inactions of State continue to perpetuate the incidence of statelessness.

The situation of Palestinians, both in the countries of the Middle East and beyond the region, illustrates strikingly the level of denial that may accompany statelessness. Palestine was a British mandate during the time of the League of Nations, up until 15 May, 1947. Under the mandate system, the inhabitants of such territories were not to be considered as nationals of the administering powers, although they might benefit from the exercise of diplomatic protection. Palestinian citizenship was regulated by statutory instrument, and included acquisition by birth, but a Palestinian citizen was not a British subject. Palestinian citizenship, as a product of the mandatory's authority, terminated with the mandate and with the proclamation of the State of Israel, even though there is some authority in international law for the continuance of certain internal laws upon the cession or abandonment of territory.

Israel had no nationality legislation until 1952. The existence of a State implies a body of nationals, and a population within a relatively well-defined territory is an accepted criterion of statehood. With one exception, however, Israeli courts held that, on the termination of the mandate, former Palestine citizens had lost their citizenship without acquiring tiny other. For the purposes of Israeli municipal law, the issue was resolved by a Supreme Court decision, and by the 1952 Nationality Law,

The Nationality Law confirmed the repeat of the Palestine Citizenship Orders 1925-42, retroactively to the day of the establishment of the State of Israel. It declared itself the exclusive law on citizenship, which was available by way of return residence, birth, and naturalization. Former Palestinian citizens of Arab origin might be incorporated in the body of Israeli citizens, provided they met certain conditions: they must have been registered under the Register of Inhabitants Ordinance on 1 March 1952; have been inhabitants of Israel on the day of entry into force of the Nationality Law (14 July 1952); and have been in Israel, or an area which became Israel, from the day of establishment of the State to the day of entry into force of the law, or have entered legally during that period These strict requirements meant that the majority of those displaced by the conflict in 1948 were effectively denied Israeli citizenship. If international law raised a presumption of entitlement to local citizenship for residents at the moment of establishment of the State, subsequent developments have made such claims redundant.

Palestinian refugees were admitted to neighbouring countries on what was expected to be a temporary basis; local citizenship, for the most part, was not available. In Lebanon, the majority of Palestinians are unable to obtain citizenship, even though prima facie some appear to qualify under the law. In these circumstances, many Palestinians appear not to be recognized as a citizen or national of any State, and may therefore be considered stateless persons.

For, whatever the political position, there is at present no Palestinian State which satisfies the international legal criteria of statehood: a permanent population; a defined territory; government; and the capacity to enter into relations with other States. There is thus no Palestinian passport entitling the holder to return to the State of Palestine, but only such travel documents as may be issued by, and allow return to, the holder's State of residericc.

Just as Israel has denied citizenship to the majority of Palestinian Arabs, the Arab countries of refugee have, for the most part, consistently rejected local integration and citizenship as a solution to a problem which, in their view, can only be resolved by repatriation and self-determination. With limited exceptions, Palestinian refugees have not been granted (and for the most part, have not sought) citizenship in the countries of refugee.

From the international law perspective, they are therefore stateless persons, notwithstanding the recognition accorded by some States to the entity 'Palestine', and notwithstanding the United Nations' recognition of the Palestine Liberation Organization as the legitimate representative of the Palestinian people. For many Palestinian citizens by birth, such citizenship will have lapsed or terminated with the events of 1948; and likewise for many, no other citizenship will have been acquired in the interim. The impact of statelessness on personal circumstances can be extensive. In the Federal Republic of Germany, a number of -judicial decisions on asylum claims have confirmed, and indeed accentuated, the disadvantaged situation of Palestinians. In a 1974 case, for example, the Federal Administrative Court ruled that a Palestinian from Lebanon could not found a claim to asylum on his flight from Israel in 1948. The applicant was stateless, and both the Constitution of the Federal Republic and the 1951 Refugees Convention provided protection only with respect to persecution feared in his land of former habitual residence.

The factual and legal situation in Lebanon changed after the Israeli invasion. From 6 June 1982, the Lebanese Government Introduced a number of measures against Palestinians those who had participated in the fighting were required to leave Lebanon, and others overseas were refused renewal of their (Lebanese-issued) travel documents. The Government stated that its objective was to reduce the number of Palestinians in the country to around 50,000. In a 1985 decision, the German Federal Administrative Court held that these measures were not dictated by '1951 Convention' reasons (that is, by considerations of race, religion, nationality, membership of a particular social group or political opinion); they were in the nature of preventive or police measures, directed against stateless persons. In the view of the

Court, residence in the Lebanon on the basis of various international agreements or undertakings could be terminated wherever the persons concerned violated national security or public order, for example, by participating in a civil war.

For the purposes of German municipal law, these rulings entail non-eligibility for asylum under the Constitution; in practice, Palestinians in the Federal Republic of Germany are 'tolerated', that is, their (international) status as refugees is informally acknowledged, so far as they are allowed to remain, although without formal status. From an international law perspective, the lawfulness of the severance of the link between Palestinians and the Lebanon is far from clear. But the theoretical niceties are of less immediate concern than the realities of prejudice and disadvantage following from the denial of national status.

Many countries have legislation permitting denaturalization. A provision of Article 66 of the 1982 Turkish Constitution, for example, declares that no Turk shall be stripped of citizenship unless an act incompatible with attachment to the motherland has been committed. However, it has been reported that some 13,788 Turks abroad have been stripped of their citizenship since 1980.

In many Eastern European countries, until the recent democratization process, loss of citizenship frequently followed illegal departure or failure to return from abroad; and was sometimes a condition attached to permission to emigrate, as in the case of Jews leaving the Soviet Union. Nowadays, governments in the region increasingly recognize the benefits to be gained from permitting greater mobility; revisions of the passport regulations were introduced in Poland, for example, specifically to encourage retention of citizenship. Poles residing abroad may now extend the validity of their passports to ten years, and an 'amnesty' was declared for many holding expired passports, who wanted to return.

Statelessness is also a by-product of refugee movements, striking at the children of those who flee. In Central America, for example, children born in refugee camps are often not registered, both because the asylum State was unwilling to contemplate their acquiring local nationality under otherwise applicable principles of jus soli, and because normal juridical relations with the parents' country of origin were In suspense. If repatriation is to a viable solution, this issue must clearly be settled, and a provision to that effect was included in the Concerted Plan of Action, adopted at the 1989 Conference. Programmes for returnees will aim,

> at regularizing the situation of returnees with regard to the delivery of identity documents and the registry of births, marriages and deaths, and other events occuring in the country of asylum and relating to the civil status of the individual. They also provide for access to citizenship for children of returnees born abroad as well as for foreign spouses....

Needless to say, these provisions do little more than reflect basic human rights principles, such as the right to a nationality in article 15 of the 1948 Universal Declaration of Human Rights, article 24(3) of the 1966 Covenant on Civil and Political Rights, article 5 of the 1965 Convention on the Elimination of All Forms of Racial Discrimination, or articles 7 and 8 of the 1989 Convention on the Rights of the Child. Moreover, article 20 of the 1969 American Convention on Human Rights is a particularly striking example of a provision that, if applied broadly and generously, would resolve many instances of statelessness:

(1) Every person has the right to a nationality.

(2) Every person has the right to the nationality of the state in whose territory he was born if he does not have the right to any other nationality.

(3) No one shall be arbitrarily deprived of his nationality or the right to change it,

There is some evidence to suggest growing recognition of the fundamental and continuing quality of a State's responsibility towards those who hold its citizenship, or otherwise possess an effective link in its regard. In 1987, for example, the United States changed its policy with respect to the involuntary return of those who had renounced U.S. nationality without first obtaining another, or without being accepted for permanent residence by another country, and who would thereby become stateless. The change in policy was prospective only, and the right to refuse re-admission in particular cases was expressly retained, The human rights dimension was not cited as a reason for the policy, however.

By contrast, the 1983 Council of Europe recommendation of stateless nomads and nomads of undetermined nationality expressly recognized both the legal problems and the humanitarian aspects of their situation." It recommended that States refrain from any measures that might lead to discrimination against nomads, encouraged the establishment of links between States and nomads, persmission to reside, travel and return, and the facilitation of family reunion.

States and international agencies know very little about stateless persons. There are few statistics, and statelessness and stateless persons are very much on the periphery, if not beyond the pale, of international organization.

The irony is that refugees and stateless persons once

walked hand in hand, and after the First World War, their numbers and condition were almost coterminous, Later, their paths diverged, with refugees being identified principally by reference to the reasons for flight, and their statelessness, if it existed, being seen as incidental to the primary cause. For stateless persons, this meant decreasing attention, accompanied by a constrained analysis of the essential characterstics of their situation. To many, statelessness was nothing but a technical problem, either the unfortunate result of a clash of municipal laws, or a divergence, or insufficient overlapping between the two traditional methods for the transmission of nationality, birth in the territory of a State, and descent. The technical side, with its prospects of harmonization of laws and có-ordination or rules, may well have lacked the fashionable touch; the tragedy is that few commentators or practitioners ever looked beyond, to the underlying human rights issues. For the central thesis -of this paper is that statelessness is indeed a broad human rights issue, even as it retains a distinct technical dimension.

Part of the problem lies certainly in the association established between stateless persons and a very restricted conception of protection.

Protection is a term of art in international law. Considered as the right of the State to exercise diplomatic protection with respect to the interests of individuals, it is inextricably bound up with the bond of nationality. That essential linkage itself reflects the principle of domestic jurisdiction, which traditionally bars other States from intervening on behalf of those who either are not its citizens or, more particularly, are the citizens of the State whose actions are impugned. This approach is characterstic of that international law doctrine which 'recognizes only States as subjects . . . . and individuals at best as . . . objects.' In such a system, once the bond of nationality is denied all possibility of protection is lost.In the Nationality Decrees Case, the Permanent Court of International Justice observed that questions of nationality are, in principle, within the reserved domain of jurisdiction. However, any impression that this entails an area of absolute, sovereign discretion is readily dispelled on even a cursory examination of international laws and State practice.

In the Nottebohm Case, the International Court of Justice emphasized in striking fashion that, if nationality rules are to he entitled to recognition in the diplomatic protection context, they must accord with the individual's genuine connection with the State. 'Nationality,' said the Court, 'is a legal bond having as its basis a social fact of attachment, a genuine connection of existence, interests and sentiments, together with the existence of reciprocal rights and duties.'

This case was concerned with one issue in particular, namely, the right of the State to exercise protection with respect to one on whom citizenship had been conferred. But the judgment can stand for more; given the existence of the juridicial facts invoked by the Court, (social fact of attachment, genuine connection, etc.), the State may be considered bound by other incidental obligations, such as that of non-expulsion, readmission, and a certain standard of treatment. It is argued below that the inherent relations linking population, territory and jurisdiction, considered together with the full range of human rights developments, now indeed imposes these and other obligations upon States.

Both the International Court and leading jurists have recognized the fundamental importance of the relationship between people and territory, and the ensuing implications for sovereignty and the responsibility of the State. In his separate opinion in the

Western Sahara Case. Judge Ammoun contrasted to its disadvantage the materialistic concept of let with a spiritual notion; the ancestral tic between the land . . . and the man who was horn therefrom, remains attached thereto, and must one day return thither to be united with his ancestors. This link is the basis of the ownership of the soil. . . .' Besides the self-determination context, similar views have been expressed with respect to the succession of States and the criteria of statehood itself; in the realm of municipal law, such linking factors are recognized as the foundation for legitimate, constitutional authority.

Brownlie notes that the relation of nationality has not been elaborated beyond the rules of diplomatic protection. International Law, it is suggested, requires the presence of certain social links, not only for exercise of the right to protection; but also for the purposes of conferment, renunciation-and deprivation of nationality. Only in the presence of such links will the purported exercise of State authority be entitled to recognition. 61 He also makes the essential point, and it needs to be made regularly, that the individual in recent times has repeatedly failed to conform to, or follow, the 'classical legal model'. And the reasons, more often than not, have been 'mobility of population, economic stress, and inequalities and political divisions which cut across international boundaries,' or the 'numerous incongruities between national laws', which produce statelessness and multiple nationality, or the political strife that results in deprivation of nationality as a measure of security.

Traditional commentators have identified two functions only for nationality: entitlement to exercise diplomatic protection; and the obligation to readmit. Writing in 1954, Scelle would have regularized the situation of the stateless unprotected by linking them directly to the international order, conferring an 'international nationality', together with 'protection' to be supplied by a special or international organ. The UNHCR provides an obvious model, but any model will fail unless built firmly on a modem conception of protection. No longer can protection be considered as exclusively incidental to nationality; instead, it is a function that attaches directly to the inherent dignity and integrity of the human personality, the totality of which is the province of the community of nations, Denaturalization and statelessness are not, or not exclusively, about the right to enter or return to a State; they are about denials of the broad spectrum of rights that flow from an individual's belonging to a particular territorial community.

The seventeenth century idea of civil society, particularly Locke's was founded upon obedience to a common law and judicature deriving from tacit consent; allegiance, not nationality, prevailed. The true end of government - dispensing justice and deciding rights by promulgated laws rather than absolute, arbitrary power - together with the promotion of the good of the people," accounted perhaps for the disregard of the narrow badge of citizenship when it came to fundamentals. In later decades, Bills of Rights and human rights could be seen to work more from the particular to the general, founding their notion of an individual's claims against the State upon the inalienability of certain rights; here, there is no place for substantive distinctions arbitrarily drawn between citizens and others. For the most part independent of citizenship, fundamental rights in the strong sense have the and-utilitarian consequence that they may not be denied by the government, even in the general interest.

The 1948 Universal Declaration of Human Rights opens in article I with the affirmative, 'All human beings are born free and equal in dignity and rights.' The

Preamble to the Charter of the United Nations proclaims the determination 'to reaffirm faith in fundamental human rights, in the dignity and worth of the human person.' The Annex to the ILO Constitution affirms that 'all human beings irrespective of race, creed or sex, have the right to purpose both their material and dignity, of economic security and of equal opportunity.' And as Sieghart observes," it is the recognition that all human beings differ from each other, and that each individual is unique, which underlies the concept of the integrity and dignity of the individual person which human rights law is primarily concerned to protect.

Inalienable rights, inherent in the individual, have often had to yield to sovereignty, the absolute assertion of right and power in a society of competing nation-States. The 1961 Convention on the Reduction of Statelessness and the comparable efforts undertaken by the League of Nations bear witness to the persistence of such claims.

Statelessness includes but it is not limited to the formal question, nationality or none. But where statelessness exists, so too does lack of protection, either across a broad or a narrow spectrum of human rights. Where there is statelessness, so there will generally be a State or States avoiding or denying one or more of the responsibilities that derive from the social facts of existence. In a primitive society of competing nation-States, the narrow badge of attribution plays its exaggerated role. As that society develops, becomes more inter-dependent, discovers a growing commonality of value and interest, the claims of the dispossessed are more likely to be heard. Hitherto, international co-operation in improving the status of stateless persons, or in reducing or eliminating statelessness, has enjoyed limited success, often disappearing down relatively unproductive paths. The time has come for a revision of such arrangements, deconstruction of earlier analytical approaches, and their substitution with practical arrangements which reflect the principles generally shared by the community of nations.

First, the constituency must be known. The United Nations, through the Commission on Human Rights and the Sub Commission on the Prevention of Discrimination and the Protection of Minorities, should be mandated forthwith to carry out a survey of the extent of statelessness worldwide; the work must not be circumscribed, either by a legal or a narrow vision, but should he as sensitive to the discrete variations in national laws and practices, as to the multiple ways by which individuals and groups may be excluded from society.

Secondly, regional initiatives at the legal level should be encouraged, for example, within the workplans of UN regional commissions. Their task should indeed be to obtain and disseminate detailed information on national laws and practices, and particularly on the ways in which they may differ, even among close neighbours. The regional dimention is important, by reason of frequently close connection between migration and loss of citizenship.

Thirdly, the lawfulness of deprivation of citizenship should be the subject of close attention, account being taken of developments in the protection of human rights at both regional and universal levels. Coupled with the right to return to the State of which one is a national, the subject may lend itself to further and better protection by way of declaration, convention or protocol.

Fourthly, and again initially at regional level, steps should be taken to establish common understandings with respect to the principles that might reasonably govern the acquisition of nationality in modern society, with a view to stopping statelessness at source, A further and related objective would be to translate the criteria of effective nationality into practical arrangements, whereby the naturalization of those sufficiently attached to the territorial State might be facilitated.

Finally, the effective protection of the rights of stateless persons must be entrusted to an appropriate international agency, in a mandate that goes beyond the essentially reactive and nonfunctional tasks so far conferred upon UNHCR:

If there is room for optimism today, it is because of renewed recognition of the genesis of State authority, which is in the people; and because the nature of State authority is beginning to be reconstructed, rather as a privilege than a right. Realistically, one may expect the society of nation-States to continue, and with it the correspondingly important indices of connection and entitlement. It should be possible, nonetheless, to envisage and encourage practical action along the lines suggested above, to eliminate the technical incidence of disconnection, to give practical meaning to the broadest range of human rights, and to allow the stateless, those who lack and are denied the legal power, to change their situation.

In the early 1990s, Germany went through a difficult debate about changes in its generous asylum laws. Much more dramatic, however, were the increases in the number of violent attacks against foreigners (and also some Germans), mainly by alienated young males. The paper discusses both, the asylum migration and the ethnocentric violence, and also possible connections between the two. In the first part, we present a typology of the perpetrators and various theories which try to explain the new violence, such as: Nazi revival, modernization theory, a theory of civilizational crisis, political culture and dominance culture, and hegemonic masculinity/masculinity crisis. In the second part, we discuss the problems of mass migration and asylum applications, and in particular the debate about it. This debate has contributed indirectly to the violence, although it is not its root cause: through open racism, through scapegoating, and also through a polarization between the two major political camps. In this polarization, different ideological traditions have led to the denial of new dimensions of social reality: the reality of multicultural ism as well as the necessity of limiting and regulating migration. In the final section, we offer a number of suggestions as how to deal with both immigration and violence against foreigners.

Migration has become a major issue in recent years for those concerned about international security or violence and peace. A growing body of literature discusses the new immigration in Europe and the reactions from governments and political groupings, especially right-wing parties and movements.' Other works have looked more closely into a particularly worrisome dimension of the interaction between nationals and foreigners: the increased and often spectacular incidence of racist personal violence.

Hostility and violence against foreigners is a phenomenon seen in many countries, but the German case, even if not a special one, will always be regarded with concern, because of the historical legacy confronting Germany. We want to do three things in this article: (1) summarize research on the character and motivations of the perpetrators, (2) present some of the major theories about anti-foreigner violence in Germany, and integrate them into a model, and (3) discuss at some length the link between immigration and anti-foreigner violence.

While this link has become an obvious controversial

political subject, very little systematic research has been done about it. Björgo mentions 'a polarised political situation in which immigrants and immigration politics are hotly debated' as one of nine main factors constituting circumstances favourable to the emergence of such violence. Björgo himself, and most other researchers on antiforeigner violence, focus on the role of government policies and the negative signals which these convey about immigration and immigrants as 'the problem'. We will argue that immigration is also a challenge for the Left and for those who usually study racism, and that at least the German Left has played a part in the polarization.

In Germany, the late 1980s and early 1990s saw a dramatic increase both in the number of people seeking asylum and in the number of attacks on foreigners. Following the new law restricting asylum, figures for both have dropped again. The view that there is a connection between the number of asylum-seekers and the violence against migrants is one held in both halves of the established political spectrum. The most concrete manifestation of a link was provided by the Hoyerswerda and Rostock-Lichtenhagen attackers and the so-called 'normal citizens' who stood by and applauded them; and a certain section of the élite was only too ready to swallow the idea. After all—to put it in Hans Magnus Enzensberger's acerbic terms—you can't expect ordinary people having to face such a huge scale of immigrants residing in their neighbourhood to accept that burning children, strictly speaking, is not permissible.

The Chancellor declared that the unreasonable requirements resulting from immigration, not the 'unreasonable state of affairs resulting from the presence of people engaging in individual or organized man-hunts', constituted grounds for a possible state of emergency; the Minister of the Interior and other politicians in government expressed concern about Germany's reputation abroad. Only gradually was it realized that, quite apart from the question of Germany's image or of the asylum problem, the failure of the state to assert its power against violent right-wing tendencies represented a threat to the very existence of the Republic.

The conservative interpretation of the link between the asylum issue and violence against foreigners could be summed up as follows: Germany has an immigration problem but not an indigenous problem with violence - or, alternatively, hostility to foreigners is a product of a dramatic influx of immigrants. Within the other half of the political spectrum, meanwhile, it has repeatedly been argued that it was only after the asylum issue was dramatized by politicians and the press that the attacks on hostels and asylum-seekers began. In a nutshell, therefore, the opposing thesis runs: Germany has a problem with violence/neo-Fascism, but no fundamental immigration problems.

We believe both interpretations are false. Both have more to do with German history than with the reality of present-day Germany; and it is probable that both actually contribute to the violence—through their own specific ways of denying reality, and through reciprocal polarization. The thesis that Germany has an immigration problem but does not, in principle, have a problem with violence - leaving aside the moral affront such a view constitutes in regard to the victims - can clearly be seen to be inadequate at the first empirical glance. Attacks and other criminal offences are by no means directed solely against asylum-seekers, but also against German-born daughters, sons, and grandchildren of immigrants who have been residents for years and are de facto members of the indigenous working population; they are directed against tourists, Aussiedler (remigrants whose German ancestors settled in Eastern Europe or Russia, or Germans from formerly German provinces in the East), disabled people, the homeless, homosexuals, and people who hold different political views.

Another reason why immigration cannot be the sole cause of the increase in acts of violence is the fact that hostility to foreigners also manifests itself in places where there is no real experience of 'foreigners', or where such experience is almost impossible. And even when situations of competition can be shown to exist, or when problems do arise from direct contact between cultures, the question still remains of why the response should take the form of acts of violence.

We nonetheless start from the assumption that Germany does indeed have an immigration problem - or at least did have one. Even if there were no anti-foreigner violence in Germany, even if there were no overtly xenophobic rightwing extremism and no covertly xenophobic sections among the population, a drastic increase in immigration, including the high number of applications for asylum, would still constitute a political and social problem. Even under more favourable conditions, a debate would still have to have been conducted about limiting immigration and applications for asylum. Having to conduct it in a context of arson and murder has been extremely difficult. Yet, as Freimut Duve, an SPD member of the Bundestag (the German parliament), stated in a very penetrating commentary on the debate about the new asylum law, the murders were not able, and were not permitted, to put a taboo on the discussion.

In order to provide detailed corroboration for our assumptions, and to develop a more nuanced account of the link between the asylum debate and violence against foreigners, we shall begin by describing the different groups of offenders and by developing an explanatory model to account for the new violence. In a further section, we address the problem of immigration and asylum, and discuss the repercussions of the political debate on society. In the final section, we present the practical conclusions that can be drawn from our analysis.

Willems et al identify four types of offenders with differing political attitudes and approaches to violence. Thus the hanger-on type does not have a pronounced right-wing or extreme rightwing world-view, and is not determinedly antiforeigner. Participation in acts of violence against foreigners is here a group-induced effect. The hanger-on is sometimes a member of skinhead or Fascho youth groups, but more often simply belongs to groups of youths who get together in their spare time. Because of his basic social attitude, the hanger-on allows himself to be drawn into anti-foreigner activities. Major factors in this connection are pressure to conform, the compulsion to solidarity, and a feeling of comradeship - but also the desire not to be ousted from the group.

The criminal youth (thug type) is generally somewhat older than the others and, at the time of the incident, already has a lengthy criminal record covering acts ranging from offences against property to grievous bodily harm. Typical features of his negative personal and professional career are: unfinished schooling; unemployment; family problems (alcoholic parents, divorce, experience of violence). These young people are action-oriented, aggressive, and inclined to violence. As a rule, there is no markedly extreme right-wing political worldview lurking behind this inclination to violence, although anti-foreigner slogans are much in use within this group. They apparently serve to compensate for experiences of failure and to firm up the individual's own identity. The diffuse violence of this type of offender is by no means directed solely against 'foreigners'; it is aimed at any kind of victim - mostly people who rank as weak, or are perceived as such.

The xenophobe or ethnocentrist is to be found mostly in the 'skinhead', 'hooligan', or 'Fascho' subcultures, but also figures in the various 'spare-time' gangs. Again, these offenders have no hard-and-fast extreme right-wing attitudes, and do not belong to extreme right-wing groups or parties. Even though their xenophobic slogans and nationalist catchwords coincide to some extent with those of the organized extreme right, they cannot simply be lumped in with the latter. Their hostility to foreigners does not derive from any extreme right-wing ideology- it results rather from a diffuse feeling that 'Germans' are suffering disadvantage vis-a-vis 'foreigners', particularly asylum-seekers. In the case of this group, worries about the future and competition for scarce goods such as jobs, housing, and welfare benefits play an important role.

Finally, in the case of politically motivated, extreme right-wing or radical right-wing offenders, anti-foreigner violence is indeed the product of an extreme right-wing ideology. Offenders of this type have contact with, or are members of, extreme right-wing parties and groups; as a rule, they have also already been politically active in these circles. The following is an extract from a judgement: The accused also took an interest in the radical rightwing groupings that existed at that time. Thus he had links with the 'National Front' (NF) which, according to the accused's testimony, is an autonomous gang that engages in clandestine activities and aims to bring about a coup on Federal territory. The accused also took part as an usher in the organization of an electoral meeting of the DVU. In addition he had personal links with the NPD.

These types of offenders attempt to act as agitators during riots, and to influence young people or direct the action - though mostly without success. The study suggests that they have only rarely succeeded in actually taking over the initiative, or even being accepted as leaders, in non-party youth groupings and subcultures. .

In the debate about the causes of increasing attacks on groups of people marked out by offenders as foreign or odd, there is a wide range of differing theories or theorems on offer. It is generally agreed that these developments cannot be explained in terms of a single cause. The unofficial view that acts of violence are the clearly definable work of isolated criminal or misguided outsiders finds little support. The discussion at the social-science level generally proceeds on the assumption that the violence is an expression of critical trends within society as a whole.

One of the international mass media's first reactions to the violence in Hoyerswerda and Lichtenhagen was: Fascism is coming to life again in Germany. The French newspaper Le Monde, for example, published a cartoon of a large swastika made up entirely of tiny matchstick people. Behind this kind of explanatory model there often lies the notion of a persistence of Fascist structures in German society, temporarily masked - for example by favourable economic conditions -but remaining present in latent form, in line with the quotation from Brecht: 'This womb is fruitful still'. The explanation for an allegedly specifically German inclination to Fascism takes two forms: one accounts for it in terms of the peculiarities of German history, the German ,special path' into the modem world; the other derives it from German characteristics, from the German mentality.

Key themes in the 'German special path' thesis are: failure of the Peasants' War in the 16th century and of the liberal-democratic revolution in the 19th century; late development of the middle class; and late achievement of German unity and nation-statehood. One feature regarded as corroborating this derivation is the fact that Fascism also rose to power in Italy, the other European state where national unity came very late, whereas in France and England, where bourgeois society emerged early on and where the nation-state was established at an early date, Fascist movements had no success. The second explanation stresses the persistence of certain elements of mental make-up, in other words specifically German characteristics such as order, cleanliness, a fixation with authority, megalomania, and militarism.

A more subtle variant of this explanatory approach is provided by recent psychoanalytical/psychohistorical work which diagnoses the continuing but latent effects of concrete historical experiences of socialization - inadequate reflection on, and processing of, traditions, notably from the time of National Socialism.

The modernization approach focuses not so much on German history as on the trends present in all industrial societies. According to this approach—as expounded, for example, by Wilhelm Heitmeyer—right-wing extremism is a reaction typical of the 'loser' in the modernization process. The accelerated pace of development in modem industrial societies, so the argument goes, produces a constant degeneration of skills, puts social rankings and professional careers in doubt, and subjects a considerable section of the population to constant worries about decline in social status. The unintelligibility of the social mechanisms by which success or failure is allotted, and the impossibility, in principle, of the individual's making any reliable plans as to the preconditions for his or her professional life career, result in a fundamental sense of insecurity. In such a climate of social anxiety, the extreme right's platform becomes a substitute for security. Its clichés used to interpret social reality offer a fourfold benefit:

— They provide simple interpretations of the social environment and the crises it manifests foreigners are taking our jobs and houses; they are the cause of the increase in crime and drug-addiction').

— They offer simple methods of removing the evil ('foreigners out!').

— They guarantee enhancement of the individual's own endangered social status ('I am a German'/'I am proud to be German').

— They provide an object of aggression that can be hated, or actually attacked, without moral scruple ('it's the foreigners' own fault; if they're persecuted, then that's only fitting punishment').

The thesis that xenophobic or extreme right-wing reactions result from collective deprivation does not presuppose any real impoverishment; the decisive factor, rather, is whether the social groups affected believe that, measured against their expectations and compared with other groups, they are disadvantaged in regard to the distribution of scarce benefits, such as job opportunities, housing, or social welfare. The frustrations produced by such comparisons then fix themselves on to the 'foreigners' who—either apparently or in reality—obtain the benefits to which these individuals believe themselves entitled.

In his analysis of right-wing extremism amongst the young, Heitmeyer underlines the effect of the shift towards individualization associated with modernization. The new opportunities for freedom which the indivdiualization of life opens up cannot be exploited by young people, and are instead, so he says, experienced as isolating and undermining. This feeling of powerlessness is reinforced by the breakdown of traditional milieus such as the family, the neighbourhood, the work-place, the trade union. This destruction of the social fabric of life, combined with the difficulty—already outlined—of formulating reliable prospects in regard to life and work in an individualized, atomized society, engender a need for clarity, certainty, and stability.

Extreme right-wing groups seemingly offer an escape from this feeling of insecurity - by providing the opportunity to identify with a group, and by having recourse to so-called ‚natural categories' such as race, nation, or masculinity. Acceptance of violence, says Heitmeyer, and the use of it, create the illusion of recaptured autonomy and power of action.

In an original essay on the subject, Eisenberg & Gronerneyer have combined elements of the classic analysis of Fascism with aspects of the modernization theory and developed them into a theory of the collapse of civilization. They distinguish three 'psycho-groups' amongst the young. The first, brought up in a traditionally authoritarian manner, has a tendency to Fascism. The second, the 'yuppie' group, exhibit more instrumental/ Machiavellian inclinations, that is to say they make their way in the world of performance and consumption using methods that range from the robust to the ruthless. However, the future, say the authors, belongs to the 'abandoned children' psycho-group, 'born amongst the concrete and later to be found stone in hand'. Their general unconnectedness is the reflection of a social and, above all, psychological/moral decivilization of bourgeois society. The violence of the future will no longer be the product of the harshness and brutality of the old 'black pedagogy' but of the increasingly widespread coldness, the lack of attention and emotional nurturing. The classic entities through which socialization took place - notably the family as an emotional refuge from society - are declining in importance, and nothing constructive is taking their place: television is replacing parents as educators. 'Consumerism' as the most recent, contradictory form of capitalism is undermining not only external nature (viz. the ecological crisis), but also the inner natures of people; morality is breaking down, inhibitions are falling away. The media embody both aspects - consumerism and brutalization. What awaits us is not another 1933 but a ghetto revolution engendered by harshness. The anger displayed by the new perpetrators of violence is not of Nazi but of narcissistic origin. And it is therefore directed, amongst others, at asylum-seekers, because they embody the possible reality of our society - real existing collapse.

The borderline area between the social and the political system is also the focus of the political culture approach. This has a number of points in common with the explanatory model based on the notion of persisting traces of Fascism and 'authoritarian character'; in this case, however, the focal point of the analysis is not just German historical traditions, but also the specific conditions governing the political culture that prevails in the Federal Republic. The term political culture is here used not in a normative or moral sense, as it is in everyday language or in the media, when there is talk of a 'decline in the political culture', meaning wrong behaviour or corruption by politicians. In the social sciences, political culture is used to describe the totality of attitudes, views, opinions, and behaviour of citizens in regard to the political system. This configuration of the subjective elements of politics determines the political climate and political consciousness of a society. It constitutes the seed-bed of social movements -and thus also of right-wing extremism.

The extent to which the criminals in Hoyerswerda, Rostock, and Hünxe saw themselves as carrying out the will of a silent majority, and the extent to which politics and the media encouraged the violence by the way in which they tackled the asylum problem - both these, for example, are matters relating to Political culture. A notable difference may be observed here between the old and new Länder. In Rostock-Lichtenhagen, the young hooligans really did believe they were simply carrying out the wishes of a large number of spectators; that they were so doing was obvious from the applause and approval of the latter. There has been nothing comparable to this in the old Länder. Here, the arson attacks have so far been staged under cover of night, with a getaway car at the ready. But this does not mean that the violence in the west has not met with tacit approval. So far, however, it has remained secret, has not dared express itself openly.

Joachim Maaz traces the difference in the political culture in the two parts of Germany to the 'second dictatorship' experience in the former GDR. A major mechanism of GDR socialization was repression by the state, by the party. The least stirring of emancipation or autonomy was strictly suppressed. This has left traces, scars in the social character. At the same time, however, inadequacies in socialization were also contained by the repression, that is to say arrested, and the system provided social security and stability in cases of good conduct. The emancipation which the breakdown of the old order has brought is thoroughly contradictory. The culture shock caused by reunification has led to a situation which we may, with Durkheim, characterize as 'anomic': it is no longer possible to live by the old norms and values, and not yet possible to do so by the new ones. And the internal assimilation of the new autonomy is taking place under (relatively) difficult external conditions. As a variation on the theme, Birgit Rommelspacher has introduced the concept of the dominance culture into the debate about right-wing extremism. She distances herself from interpretations of right-wing extremism as a way of coping with threat, and stresses that it is not primarily real problems such as impoverishment or social deprivation that lie behind the activation of extreme right-wing tendencies. What is involved, she claims, is rather the safeguarding of privileges and, where it comes to conflict, a process of 'expansive empowerment'. In the former GDR, says Rommelspacher, there is a 'racism born of weakness'. In order to cover up the shame provoked by self-devaluation, individuals identify that much more strongly with the nation. The new nationalism in the western Länder, meanwhile, feeds on opposite kinds of motives. The establishment of the new political order brought a considerable narcissistic boost to the former FRG. The other side's defeat reflected the superiority of the Federal Republic's own political order and economic might. In the 'delirium of self-satisfaction', this superiority has constantly to be reconfirmed; disturbances in self-image produced - so an extrapolated interpretation might run - by disillusionment about the costs of unity and other critical developments, such as rising unemployment, have to be compensated for in a 'cycle of dominance' involving the devaluation of third parties.

The predominance of men amongst offenders - and also, incidentally, amongst the victims of violence - has prompted various authors to see youth violence as an expression of a masculinity cult or crisis of masculinity. Lothar Böhnisch and Reinhard Winter, for example, view extreme right-wing behaviour, with its high degree of acceptance of violence and its xenophobia, as displaying parallels with basic structures in the masculine approach to coping with life, being simply an exaggerated form of this. In a patriarchy, men are integrated into a hegemonial culture which appears to be theirs, but which ultimately is not. The price for the power deriving from being a man is denial of access to one's own feelings and needs, to one's 'true self', which is distorted with attributes of external 'masculinity'. The debasement of the female and of the weaker men is a stereotypical feature of this kind of shell-like masculinity. Where the basic structure of all-male clique-style groupings amongst the young develops within the context of daily experience of social and economic powerlessness, it constitutes - so Böhnisch & Winter argue - a fertile breeding-ground for extreme right-wing tendencies.

A variant of this theorem of instability in masculine status can be derived from the detailed empirical study of male milieus by Walter Hollstein. In his investigation, Hollstein starts from the assumption that the actual lives and experiences of men, particularly sexual relations, are undergoing major changes. These changes are reflected in the self-images of the men questioned. According to Hollstein they signal a 'momentous change in masculinity', albeit with notable variations according to social class: the image of traditional masculinity, with its hardness, competitiveness, and conflict, is being defended by a 'curious coalition' of men from the upper and lower classes; the greatest potential for change, meanwhile, lies with the two middle classes. The lowest level of the lower class is 'the machos' last but still extremely sturdy bastion'. Only in the lower classes does one still find unabashed pride in masculinity, but at the same time this group of men is the least satisfied in regard to the fulfilment of its life's goals. This double discrepancy between the established but outmoded male self-image and the changing/unsatisfactory external male world as diagnosed by Hollstein may constitute an important foundation for right-wing extremism and for violence. Both would be a means of compensating for crumbling masculinity.

Joachim Kersten also establishes a link between everyday patriarchy and male youth violence. Like Birgit Rommelspacher, he distances himself in his analysis from attributions of cause involving disintegration and lack of direction. Instead, he follows Robert W. Connell in seeing male youth violence as an expression of conflicts within 'hegemonial masculinity'. In this interpretation, the two aspects - predominantly male aberrant behaviour on the one hand and its control on the other - are seen as cultural practices for regulating the legitimate and illegitimate use of violence within public and private male domains. 'Good' hegemonial power extends to sexual precedence over women and the production of offspring, to dominance in areas where there is a legitimate use of violence as a means of controlling other (hostile) masculinity, and to the provision of material necessities/control over production. The perverted masculine self-image that prevails amongst neighbourhood gangs, 'skins', and 'hooligans' falls back, so Kersten argues, in an altered form, on exactly the same functions of quasi-legitimate combativeness. Youth violence is thus ultimately characterizable as 'male-dominated aberrant behaviour' and contention over 'hegemonial masculinity'.

We do not wish, at this point, to put for-ward an alternative theory of our own. We believe the existing range of theories is quite sufficient and that, despite one or two internal contradictions, it is possible to tie them in with one another. The only reservation we have - not least because of the profile that emerges of the offenders - concerns those attempts at explanation that resort immediately to theories about Fascism or to Germany's special path. Even if one takes the political/ideological and organizational background of the New Right and extreme right wing in Germany seriously, one also has to take account of the historical changes that have occurred in the Federal Republic since World War 11, and of similar problems in the domain of right-wing radicalism or hostility to foreigners in other countries. Racist violence is a serious problem throughout Europe - 'from Sweden to Italy, from Britain to Russia', and even countries with long liberal-democratic traditions, such as Sweden or The Netherlands, have seen an upsurge in violent offences against migrants and foreigners .8 Recourse to the familiar and easily comprehensible interpretative model based on Fascism and, in particular, German National Socialism, may distort analysis of the new violence rather than clarify it - even if we do continue to take into account the persisting effect of cultural and in particular psycho-historical factors.

If we make a distinction between, on the one hand, the socio-psychological causes of the inclination to violence of what are mainly male youths, and, on the other, the direction in which this violence is aimed, it becomes clear that the theorems described above complement rather than exclude one another.

An integrated model might then look as shown in Figure 1. According to this picture, in overall social terms (we here exclude specific individual and family background) it is primarily the social situation (e.g. anomie, the problems associated with modernization) that generates anxieties and anger. The fact that these anxieties and this anger are then translated into violence - and not into socially acceptable forms of, for example, political activity - would be explicable primarily in terms of the 'collapse of civilization' theorem and the theorems about masculinity (crisis of masculinity, hegemonial masculinity). As regards the direction taken by this violence, political culture, dominance culture, and right-wing extremism then play an important role. It is within this context that the debate about immigration has its locus in terms of the system as a whole. The overt and covert messages conveyed in this debate help to determine against whom the violence is directed, even if they are not a direct cause of it.

Official reaction to the new 1992/3 law on asylum followed general political predispositions: sighs of relief from the ruling parties and certain sections of the SPD; responses ranging from disappointment to outrage on the Left and amongst refugee organizations. 'The law on asylum lies in ruins, and with it the little bit of human dignity guaranteed by Article I and the little bit of legal recourse guaranteed by Article 19', wrote Viktor Pfaff, one of the leading lawyers involved in legal matters relating to foreigners and asylum in the Federal Republic, in an analysis for a leftist magazine. The heavyweight liberal press was not stinting in its criticism either. There was not much left, it said, of what was once the most liberal asylum legislation in the world. The compromise over asylum was founded on a tacit agreement that the asylum problem should be passed on to neighbouring countries. That the old claim that the right to asylum must in principle be granted to all victims of political persecution, and be granted in Germany, was no longer tenable, allright; but how had it been possible for this to have turned into the thesis that victims of political persecution must enjoy the right of asylum - as long as it was not in Germany? Would it not have been more honest simply to abandon the idea of the basic right to asylum as a subjective individual right?

Whatever the political evaluation, any social-scientific analysis will have to begin with the almost tautological observation that the kind of 'compromise on asylum' that has emerged is one that corresponds to the social balance of power and the ideological and party-political configuration in the Federal Republic as it existed at the time. All parties with political responsibility were under pressure to limit immigration. They also wanted to maintain the fiction, however, that they were still prepared to help genuine victims of persecution, even under the radically altered conditions. The decision as to whether Germany needs any immigration, and, if so, how much, is one that politicians have not yet confronted. And as far as the decision about how many refugees Germany was prepared to accept in future, no agreement was possible - though it must also be said that the old constitutional obligations prevented such a decision from being arrived at freely. The new law on asylum avoids this issue: possible avenues to a hypothetical decision-making process such as this have been drastically curtailed.

But what was the alternative? Even an enlightened discussion about asylum is difficult because the problem is embedded in other controversial issues associated with immigration: the status of foreigners who have settled in Germany, and of those generations of immigrant families who were born here-, the return of descendants of German settlers from eastern Europe and the CIS; illegal immigration; refugees from civil war. And the precise relationship of the asylum question to these topics is, in its turn, a disputed one.

Experts on migration agree that the Federal Republic has long since become an immigration country. The thesis that Germany is again in need of immigration in the medium term, for economic and demographic reasons, is also broadly established. However, as is shown by a glance at those states that view themselves as immigration countries, and pursue a policy of immigration, this kind of directed immigration would, in all probability, be highly selective, and indeed might have counterproductive effects (brain dram') from the point of view of development for the emigration countries. In view of the recession and growing mass unemployment, there is at present no likelihood of mustering any political majorities in favour of regulated labour migration into Germany - other than for seasonal or contractual work. All the more so since Aussiedler will continue for some years to encroach on the labour market, although their influx has now been brought under control through the imposition of limits on the numbers accepted annually. On top of this there are illegal immigrants - in 1992 these are reputed to have numbered 300,000 - who are absorbed (and exploited) by the black economy and some of whom also become caught up in, or dependent on, criminal activities.

Because of the continuing crisis in the Third World, and because of the end of the East-West conflict, migratory movements have acquired a new historical quality; they are becoming a prime area of conflict in international relations. As a consequence of the economic divide, and as a result of asymmetrical population growth, pressures to emigrate to the well-off industrialized countries are growing in the developing South, which has now been enlarged by the addition of various parts of the former Second World. The basic conflict consists in the fact that more people want to immigrate than the immigration countries are willing or able to take in. This, incidentally, applies across the whole world, with few exceptions: the number of states that feel overburdened by immigration is on the increase.

The only ultimately effective strategy and constructive alternative to an unsocial sealing off process that comes up frequently in discussions is that of tackling the causes of migratory movements. The industrial countries undoubtedly contribute in many ways to current migratory and fugitive movements - through the balance of power in the world economy, through the privileged exploitation of resources or by charging up regional conflicts through arms supplies. But even if there were a fundamental change of course and maximum political commitment, no more than marginal rectifications to the causes of migration could be expected.

The problem affects both sides - the societies undergoing immigration and the immigrants themselves. Using the German example, Horst Afheldt has put forward impressive arguments to show that, given the current structural defects in economic development, a continuing high rate of immigration would present not only a political/psychological but also a material threat to democracy and the welfare state.

Fear of foreigners, and hostility to foreigners, do not manifest themselves only, or even primarily, as ideological but as social phenomena. The migrants are not just foreigners, but poor foreigners, in search of housing, jobs, social welfare, prospects for themselves and their children. Here they come up against people belonging to the lower third or quarter of society, who defend this standard of living by virtue of nation-based membership of this society. This is exactly the point at which the ideologization of the problem, and radical right-wing agitation, begin.

What does all this have to do with the asylum issue? Shouldn't a liberal and wealthy country like the Federal Republic, though it does not wish to be a country of immigration or at least does not want any further immigration at present, keep its doors open to people fleeing to Germany in order to escape political persecution, war, or civil war? Again, this question is no longer as easy to answer as it was at the time of the East-West conflict. The 'global utopia of former Article 16' (Freimut Duve) stood for thirty years under the dubious aegis of the Wall. And even then, Art. 16, para. 2 was not practised as universally as was implied by the constitution. Thus in 1986 - to quote just one example - the Federal Republic successfully approached Erich Honecker, the General Secretary of the SED, with a demand that the gap which the GDR had opened up in the said Wall, in order to allow through victims of political persecution in Iran, be closed up again. From the end of the East-West conflict, the law on asylum - in the absence of other legal immigration options - became a signal inviting economic refugees from all over eastern and south-eastern Europe to immigration.

To what extent the old German law on asylum, with what were in principle and by comparison very generous provisions, was actually 'abused' in order to secure immigration without the necessary grounds of political persecution, is a disputed point; and the 'abuse' is itself abused for political purposes. That the acceptance level for asylum applications is so low in Germany - it has declined noticeably in all the countries of Europe over the last few years -also has to do with the fact that 'political persecution' was defined ever more restrictively, even before the reform.

The great majority of asylum-seekers come from countries where there is war or civil war, where ethnic minorities are discriminated against, or individuals/groups suffer political persecution. This remains the case even though in many instances - especially in Germany - an economic motive can be imputed for the request for asylum. The fact that asylum-seekers also make their way into the Federal Republic with the help of illegal organizations cannot be taken as automatically indicating 'abuse'. If someone is being persecuted, the help of an organization such as this can be their salvation. On the other hand, it is clear that the illegal operators engage in a constant cat-and-mouse game with the German authorities in an attempt to circumvent the rules governing the asylum procedure, and generally do not care about the status of their clients or even their lives.

But the attempt to distinguish between ‚abuse' and 'use' of the asylum law distorts perception of the true problem. It is, for example, morally thoroughly dubious to refuse someone asylum because she or he is not being politically persecuted but is trying to flee intolerable living conditions - i.e. structural violence - simply on the grounds that this kind of violence kills them more slowly. This puts advocates of a liberal asylum policy before a dilemma. As critics of restrictions on immigration, they rightly point out that even the old law on asylum was subject to considerable limitations, that many asylumseekers leave the country again, and that a country such as the Federal Republic would have no trouble at all coping with what, in comparison with the total population, is a tiny volume of immigration. Thus they argue, using the still comparatively low figures of the late 1980s. On the other hand they put forward sound arguments for an extension of the grounds of acceptance and for better provision for asylumseekers. The resultant problem that Germany would then necessarily become an even more attractive country for asylum is, however, not discussed. Moreover, every analysis that used to stress the numerically insignificant burden to Germany, has, until the change in the asylum law, continually been overtaken by drastic increases in applications.

Working out a non-cynical ethical stance on the asylum issue is harder than discussing ethical questions relating to apocalyptic nuclear threat, wrote Freimut Duve in the article cited below, because any kind of restriction on asylum is inhuman. And yet there is a risk that the progressive camp will have to bid farewell to the spurious solutions with which it believed a German guarantee of an individual right to asylum for all victims of political persecution and a right of residence for the greatest possible number of refugees from civil war could be championed against a changed world situation and the partially fragmented social reality that now prevails in Germany.

In addition to the approach of tackling the causes of flight, which we have already mentioned, these spurious solutions include the 'immigration laws' proposal or the so-called 'European option'. Even if the European Union succeeded in formulating a joint policy on immigration (there will be no purely national policies in Western Europe), this would not necessarily help those trying to enter a West European country. Quotas, e.g., would be reached very quickly at the beginning of each year, so the problems of entry via the asylum route or by illegal methods would automatically arise again. The as yet non-existent collective European solution to the asylum question will neither provide Germany with any significant relief nor be geared to a high standard of immigration. All over Europe there is a marked trend towards reducing the possibilities of access and residence for refugees. In its resolutions, the executive committee of the United Nations High Commission for Refugees talks of burden-sharing in coping with mass refugee movements. As is well known, German diplomatic efforts to achieve a more even distribution of asylum applications and civil-war refugees were not very successful.

Ultimately, asylum in cases of mass immigration can be assured only if there is a method that is acceptable from both the political and humanitarian point of view of making - and above all enforcing - decisions about who may stay and who must leave. The problems associated with accelerated processing of applications have been made very obvious by the new asylum law; and so has the consequence - the deportation of thousands every year.

The real nub of the matter, however, is the question of the welfare state and the social contract. The Federal Republic of Germany could, in an emergency, accommodate and provide for millions of refugees in a relatively humane fashion. But for this to be possible, society would have to make available a significant proportion of its income, and this, it is clearly not willing to do. The left-wing and liberal sections of the middle class would lose this fight against those of their compeers, who were unwilling to relinquish any of their prosperity, and, above all, against that third of society that has nothing left to give away of what, in a German context, is a very relative prosperity. And anyway there must be great doubts as to whether this kind of commitment is at all credible. Asking for housing of refugees is a simple matter if you do not have to compete for scarce and, most importantly, affordable accommodation. The section of Federal German society that is particularly liberal-minded by no means coincides with the group of Germans who have to bear the brunt of the processes of integration required by immigration.

Sibylle Tönnies has given a succinct account of the problem raised by the controversy about the new asylum law. She expressly excludes the question of whether the large number of asylum applications was not actually very justifiable, and whether it could not have been made even greater on humanitarian grounds. Instead, she points to the fact that the state could not enter into this debate with the population on the basis of guaranteed asylum for every individual because it really did not know whether the numbers involved were legion or not. It was therefore unable, says Tönnies, to give the population that feeling of security which the latter properly demands by virtue of the social contract.

To this extent, the reproach directed at the majority of the SPD party in the Bundestag for its support of the new law, which required a two-thirds majority, has to be considerably modified. What is (and was) at issue is not just power - which the SPD still does not hold - but also democracy. The observation which even many left-wing politicians have made to the effect that widespread feelings amongst the population cannot be ignored in the long term should not be precipitately dismissed as cheap populism. What lies behind it is, after all, the essentially banal political realization that asylum policy in particular depends on a minimum of approval amongst the indigenous population.

Finally, as far as the issue of help or solidarity is concerned, even where the most progressive policy imaginable is in force, there will always be many people left who are not helped. And the question of how the resources which society is prepared to make available for the support of refugees can be used most effectively is one that is open to debate. Another of the paradoxes associated with the asylum issue is that Germany can help more refugees if fewer of them go there. This observation would be cynical only if it were used as an excuse not to intervene; in economic terms, it is almost a commonplace. To care for refugees in many other countries would cost the Federal Republic - even if it assumed all the expenses - only a fraction of what it costs to accommodate them in Germany itself. And the United Nations provides far fewer resources for supporting 'real' refugees in the Third World than is spent in western Europe on economic refugees with no claim to asylum.

As has already been indicated, we hold the view that the debate about immigration and asylum in Germany has contributed in three ways to an aggravation of the problem and has also directly or indirectly fostered the various acts of violence and other attacks against foreigners. The clearest contribution is that made by the racist rhetoric of the New Right, with ramifications reaching into the conservative end of the political spectrum. In addition, migration and asylum are used by a section of the established parties as a scapegoat to offset social, political, or educational/psychological failings. But this kind of abuse of the asylum problem is not engaged in only by the political classes; it also has roots in the population. The scapegoat mechanism is a classic power technique for offsetting loss of legitimation. At the same time it provides ways for socially disoriented or marginalized groups to compensate for their situation; and it enables less disadvantaged strata of the population to rationalize defensive attitudes.

The third factor that has helped aggravate the situation, however, stems from the polarized debate that has been going on for years between the broad political groupings, and in which they have laid the blame on each other for a situation whose proper analysis would have required them to abandon certain traditional denials of reality. These denials are propelled by powerful political fantasies or traditions - on the one hand, the nationalist, and, on the other, the internationalist tradition. As ideal-typical tendencies they are not entirely congruent with the two broad political camps, but their distribution across them varies significantly.

The national tradition is the cause of the stubborn refusal, particularly within CDU and CSU circles, to accept, or at least accommodate oneself to, the fact that Germany has long since become an immigration country.

But even within the internationalist tradition - to which, in terms of basic principles, peace researchers generally feel much closer - denials of reality may also be found. Instead of there being even any thought of entering into a discussion about the competition for scarce benefits, the rhetoric about 'the boat being full' is often simply reversed and the xenophobia is countered with a xenophilia that refuses to acknowledge the genuine social conflicts and the organised and poverty-induced crime. 14 For example, for far too long, the Left has - as the Greens' Senator for Foreigners in Bremen, Helga Trfipel, put it - suppressed the issue of how the asylum route is used to get round the inadequate immigration provisions. Further typical denials of reality are the three 'magic formulas' mentioned above: tackling the causes of flight, the European solution, and immigration legislation. We are very much in favour of political measures in these areas, but we have strong doubts as to whether these would obviate the need for limits on immigration and a restriction on global guaranteed asylum.

Social scientists of a psychological bent have repeatedly pointed out that both these tendencies - the one that makes bogeymen out of foreigners and the one that idealizes them - are linked to a disturbed relationship to one's own, and in this case also the German, identity. Might it be a case of proving who is the better German? Right-wingers are inclined to adhere to the monocultural nation-state, which no longer exists - and in the strict sense never existed, and left-wingers tend to cling to what is foreign, 'in order to have nothing more to do with the traditions in their own country.

There will probably only be a chance of overcoming the nationalist tradition if the German Left for its part addresses the real existing problems of migration and asylum and draws the appropriate conclusions, instead of 'blithely continuing to lay claim to everything good, like tolerance, humanity, and internationalism'. After all, the appeal for solidarity and multiculturality becomes implausible or ineffective if it excludes consideration of the difficulties involved in implementation, and if it does not take seriously the living conditions and anxieties of those on whose political support it remains dependent. Even enlightened interpretations of the migration and asylum problem have no chance of success with the - in principle - well-meaning majority if they do not ensure that the latter's right to live within familiar economic, social, and psychological surroundings is also taken into account in analysis and political programming.

In Germany, violent dispositions, particularly amongst male youth, have increasingly been directed against foreigners and asylum-seekers. While the trend seems to have been reversed, the number of violent offences remains very high, much higher than it used to be in the 1980s. The causes of this violence lie - in addition to the specific individual and family context of the offenders - in overall social conditions and developments: the anorme following on from the collapse of the former GDR; existing or impending modernization-induced losses in both parts of Germany; the dominance culture; general crises of civilization; and problems relating to masculinity.

German politics has failed: (a) to distance itself promptly and sufficiently clearly from the violence; (b) to conduct the debate about asylum in a way that does not reinforce hostility towards 'outsiders'. The inclination to violence, and the direction it takes, have been promoted by blatant racism and the use of scapegoat mechanisms by certain sections of both the political class and the population. But it has also been encouraged by the polarization of the asylum debate between a national tradition on the one hand, and an idealizing international tradition on the other. If this polarization, and the denials of reality associated with it, were eliminated, the following analysis of the situation would, in our opinion, secure general assent:

Germany has a new violence problem. The violence is directed especially against foreigners and asylum-seekers, but not solely against them. We cannot expect that the propensity to violence will disappear if no further immigration takes place. Even if it were possible to reduce immigration and asylum-seeking to nil - in other words, even if it were ethically defensible, politically sensible, and practically feasible -Germany would still have the problem of how Germans (or those who consider themselves exclusively as Germans) treated residents of their country who are regarded, either constitutionally or socially, as foreigners or strangers. This means, to begin with and above all, facing the challenge of safeguarding these people and other individuals at risk, against attack and discrimination.

Germany is de facto an immigration country, and has been one for a long time although it does not officially see itself as such. All the indicators in immigration research show this. It is true irrespective of whether, at present or in the near future, there is any significant amount of legal immigration additional to the admission of other family members. Indeed, in the medium term, Germany will probably be economically and demographically dependent on further immigration.

Germany has an immigration problem. More people want to come to Germany than Germany would be able to absorb even if it were so willed; in other words, immigration to Germany must be limited. Given the overlap between asylum-based and general immigration, limits on immigration in general always mean limits on asylum. The harsh truth is that, under the radically altered global political conditions, Germany cannot maintain the general guarantee of asylum even for victims of political persecution.

A number of practical consequences could then be drawn from this analysis of the situation: As far as the prevention of violence is concerned, we agree with those who consider new legislation - e.g. a tightening-up of juvenile law neither necessary nor justifiable. Apart from the legal scruples involved, this discussion once again merely distracts attention from the real problems. The most effective deterrent will probably turn out to be high-quality police-work (speedy intervention, increased surveillance of the right-wing scene) and a resultant high clear-up rate. Over and above this, we recommend increased protection for potential victims, including quite ordinary, but only seemingly trivial measures such as improved fire protection in communal accommodation.

Medium-term preventive measures could include 'antagonistic co-operation' with young people inclined towards violence and xenophobia. A precondition for this would be avoiding too hasty negative categorizations: the problems and difficulties of these young people must be taken seriously. In the domain of political education, issues and methods would have to be redefined, with an eye to right-wing extremism and xenophobia. It is not sufficient constantly to develop new teaching models and convey new cognitive matter; it will also be necessary to reflect on the conditions in which teaming takes place and about internal social/emotional structures in school. Fostering the skill needed to resolve conflicts non-violently is not primarily an intellectual task. The media have a special responsibility as regards the reporting of events and the general violence that is on offer as part of entertainment. Agreements designed to avoid commercial competition for the best violent effects and greatest violence-appeal are long overdue.

The narcissistic disturbances, lack of empathy with victims, and inadequate social behaviour to which emotional shortcomings in overall socialization lead, occur amongst many children, not just amongst youths with an overt inclination to violence. To tackle these shortcomings, there is a need, amongst other things, for a social and family policy. Under the present overall economic and social conditions (breakdowns of the classic division of tabour and of old family structures), this means first and foremost a quantitative and qualitative improvement in nursery and all-day schooling. A central element in any counterstrategy should be consideration of how the masculinity cult might be rectified, and how the experience and practice of masculinity might be influenced and politically modified.

As to relations with immigrants who have settled in Germany, and with their descendants, the symbolic integration signalled by demonstrations and chains of candle-lights must be followed up by integration of a legally based material kind. Relevant measures include the granting of naturalization after an established period of residence, making it easier to hold dual nationality, and the granting of more political rights, e.g. extending the right to vote in local elections to others besides EU citizens. In addition, the possibility of an anti-discrimination law should be discussed.

As regards immigration, we consider the imposition of limits, including limits on the right to asylum, to be indispensable. We have furnished ample arguments in support of this view. At the same time, every attempt should be made to rectify the curtailments in the rule of law that have already occurred, and to put a stop to the ‚structural violence' against foreigners that is practised by the administrative authorities. It would be deeply ironic if Germany succeeded in reducing the number of deaths among immigrants resulting from personal anti-foreigner violence to zero, only to see it replaced by an increasing number of suicides among asylumseekers resulting from that institutional violence.

In addition, we argue for greater commitment on the part of Germany to worldwide aid to refugees and for a foreign policy that is more human rights-oriented as a whole. All this would not 'solve' the problems occasioned by migratory movements - there is currently no prospect of a fundamental solution of this kind - but Germany would retain more credibility as far as the necessity to practise selectivity is concerned.

For the first three centuries or so after the European discovery of the New World the principal source of immigrants to the two American continents and the Caribbean was not Europe but Africa. Only in the early nineteenth century did the accumulated total of European settlers in the New World exceed the approximately 10 million Africans who had made the trans-Atlantic voyage in the years since 1492. To explain the African diaspora by citing entrepreneurial instincts, the love of democracy, or the freely chosen decisions of migrants to follow the lodestar of American promise would be a mockery. Clearly, the involuntary movement of those 10 million Africans is best explained not in terms of their individual characters and choices but in terms of the catastrophically disruptive expansion of largescale plantation agriculture and its accursed corollary, largescale commercial slavery.

A comparable-though, to be sure, not identical-element of involuntariness characterized emigration from nineteenth-century Europe. Any generalization about what prompted a phenomenon as long-lived and complicated as the great European migration must, of course, be subject to many qualifications. All discussions of the migration process recognize both push and pull factors. But at bottom the evidence convincingly supports the argument that disruption is essential to the movement of people on such a scale. And, as in the African case, the best, most comprehensive explanation for a process that eventually put some 35 million people in motion is to be found in two convulsively disruptive developments that lay far beyond the control of individual Europeans. Those developments had their historical dynamic within the context of European, not American, history.

The first of these needs little elaboration. It was, quite simply, population growth, In the nineteenth century the population of Europe more than doubled, from some 200 million to more than 400 million, even after about 70 million people had left Europe altogether. (Only half of these, it should be noted, went to the United States-one among many clues that the American-magnet explanation is inadequatc.) That population boom was the indispensable precondition for Europe to export people on the scale that it did. And the boom owed little to American stimulus; rather, it was a product of aspects of European historical evolution, especially improvements in diet, sanitation, and disease control.

The second development was more complex, but we know it by a familiar name: the Industrial Revolution. It includes the closely associated revolution in agricultural productivity. Wherever it occurred, the Industrial Revolution shook people loose from traditional ways of life. It made factory workers out of artisans and, even more dramatically, turned millions of rural farmers into urban wage-laborers. Most of those migrants from countryside to city, from agriculture to industry, remained within their country of origin, or at least within Europe. But in the early stages of industrialization the movement of people, like the investment of capital during the unbridled early days of industrialism, was often more than what the market could bear. In time most European societies reached a kind of equilibrium, absorbing their own workers into their own wage markets. But in the typical transitional phase some workers who had left artisanal or agricultural employments could not be reabsorbed domestically in European cities.

The large scholarly literature documenting this process might be summarized as follows: Imagine a map of Europe. Across this map a time line traces the evolution of the Industrial Revolution. From a point in the British Isles in the late eighteenth century the line crosses to the Low Countries and Germany in the early and mid nineteenth century and to eastern and southern Europe in the late nineteenth and early twentieth centuries. Across the same map a second line traces the chronological evolution of migration to the United States. As it happens, the two lines are almost precisely congruent-migration came principally from the British Isles in the eighteenth and early nineteenth centuries, then mainly from Germany, and finally from the great watersheds of the Vistula and the Danube and the mountain ranges of the Apennines and Carpathians to the south and east.

The congruence of those lines is not coincidental. Industrialization, in this view, is the root cause and the most powerful single variable explaining the timing, the scale, the geographic evolution, and the composition of the great European migration.

For another perspective on the importance of understanding the European migration from a European point of view, consider the lyrics of a nineteenth-century Italian folk song called 'The Wives of the Americans." In this case, the "Americans" were men who had gone off to America and left their wives behind in Italy-specifically, the southern region of Campania. In fact, men, young men in particular, predominated in the nineteenth-century migratory stream, and their predominance constitutes a reliable indicator of their purposes. Many of them never intended to settle permanently elsewhere but hoped to work abroad for a time and eventually return to the old country. Repatriation rates for European immigrants averaged nearly 40 percent. Only the Jews and the Irish did not go home again in significant numbers. For some later, "new" immigrant groups, especially from the southern Danube regions, repatriation rates ran as high as 80 percent.

The song describes the wives of the Americans going to church and praying, "Send money, my husband. Send more money. The money you sent earlier I have already spent. I spent it on my lover. I spent it with pleasure. Send more money, you cornuto fottuto [damnable cuckold]." Those lyrics conjure an image of immigration quite different from the one General Patton urged on his Italian-American troops in 1943. Together with the figures on repatriation, they offer a strong corrective to uncritical reliance on the American-magnet explanation for the past century's European migration.

What happened to European immigrants, and to American society, once they arrived? Much historical inquiry on this point focuses on immigrant hardship and on recurrent episodes of nativism, anti-Semitism, anti-Catholicism, and anti - foreign-radicalism, from the Know-Nothing movement of the 1850s to the American Protective Association of the late nineteenth century and the revived Ku Klux Klan of the early twentieth century, culminating in the highly restrictive immigration legislation of the 1920s. Those are important elements in the history of American immigration, and we would forget them at our peril. But getting the question right is the most challenging part of any historical investigation, and there is an analytically richer question to be asked than Why did immigrants meet sometimes nasty difficulties?

The explanation is surely not wise social policy. Beyond minimal monitoring at the ports of entry, no public policy addressed the condition of immigrants once they were cleared off Castle Garden or Ellis Island. But three specific historical circumstances, taken together, go a long way toward composing an answer to the question.

First, somewhat surprisingly, for all their numbers, immigrants-even the 17 million who arrived from 1890 to 1914-never made up a very large component of the already enormous society that was turn-of-the-century America. The census of 19 10 records the highest percentage of foreign-born people ever resident in the United States: 14.7 percent. Now, 14.7 percent is not a trivial proportion, but it is a decided minority, and relative to other societies that have received large numbers of immigrants, a small minority. The comparable figures in Australia and Canada at approximately the same time were 17 percent and more than 20 percent, and even higher in Argentina. So here is one circumstance accounting for the relative lack of social conflict surrounding immigration a century ago: at any given moment immigrants were a relatively small presence in the larger society.

A second circumstance was economic. Immigrants supplied the labor that a growing economy urgently demanded. What is more, economic growth allowed the accommodation of newcomers without forcing thorny questions of redistribution-al ways the occasion for social contest and upheaval. Here, as so often in American history, especially during the period of heavy immigration before the First World War, economic growth worked as a pre-emptive solution to potential social conflict.

The third circumstance was more complicated than sheer numbers or economic growth. I call this circumstance "pluralism"-by which mean simply that the European immigrant stream was remarkably variegated in its cultural, religious, national, and linguistic origins. These many sub currents also distributed themselves over an enormous geographic region-virtually the entire northeastern quadrant of the United States-and through several political jurisdictions. By the 1920s immigrants were distributed widely across the great industrial belt that stretched from New England through New York, New Jersey, Pennsylvania, and beyond: Ohio, Indiana, Illinois, Michigan, Wisconsin, and Minnesota. The states with the most immigrants, not incidentally, also had per capita incomes higher than the national average-an important fact pertinent to understanding the relationship between immigration and economic vitality.

The varied composition and broad dispersal of the immigrant stream carried certain crucial implications, one being that no immigrant group could realistically aspire to preserve its Old World culture intact for more than a few generations at best. To be sure, many groups made strenuous efforts to do just that. Legend to the contrary, last century's immigrants did not cast their Old World habits and languages overboard before their ship steamed into New York Harbor. In fact, many groups heroically exerted themselves to sustain their religions, tongues, and ways of life. The Catholic school system, which for a generation or two in some American cities educated nearly as many students as the public school system, eloquently testified to the commitment of some immigrant communities to resist assimilation. But circumstances weighed heavily against the success of such efforts. The virtual extinction of the parochial school system in the past generation-the empty schools and dilapidated parish buildings that litter the inner cores of the old immigrant cities-bears mute witness both to the ambition and to the ultimate failure of those efforts to maintain cultural distinctiveness.

A second and no less important implication of pluralism was that neither any single immigrant group nor immigrants as a whole could realistically mount any kind of effective challenge to the existing society's way of doing things. No single group had sufficient weight in any jurisdiction larger than a municipality to dictate a new political order. And there was little likelihood that Polish Jews and Italian Catholics and Orthodox Greeks could find a common language, much less common ground for political action.

The most comprehensive explanation of the causes of immigration a century ago is to be found in the disruptions visited on European society by population growth and the Industrial Revolution. The United States was, to use the language of the law, the incidental beneficiary of that upheaval. The swelling immigrant neighborhoods in turn-of-the-century American cities were, in effect, by products of the urbanization of Europe. And once landed in America, immigrants accommodated themselves to the larger society-not always easily assimilating, but at least working out a modus vivendi-without the kinds of conflicts that have afflicted other multinational societies. That mostly peaceful process of accommodation came about because of the relatively small numbers of immigrants at any given time, because of the health of the economy, and because of the constraints on alternatives to accommodation inherent in the plural and dispersed character of the immigrant stream.

Index